Small Group Instructor Training Course (SGITC)

Volume 2:

Training Support Packages

Small Group Instructor Training Course (SGITC)

Volume 2:

Training Support Packages

Small Group Instructor Training Course (SGITC): Volume 2: Training Support Packages

Copyright © 1998, 2003 by Integrated Development, LLC,
Copyright © 2011 by PharmaLogika, Inc.

PharmaLogika, Inc.
PO Box 461
Willow Springs, NC 27592

www.pharmalogika.com

PharmaLogika and its logos are trademarks of PharmaLogika, Inc. All other trademarks used herein are the properties of their respective owners and are used for identification purposes only.

All rights reserved. No part of the material protected by this copyright notice may be reproduced or utilized in any form by any means, electronic or mechanical, including photocopying, recording, or by any information storage and retrieval system, without the prior written permission of the copyright holder.

The user is put on notice that while we have endeavored to reproduce the text provided by the government faithfully, mistakes may exist and that as such the company disclaims any responsibility for anomalies that may be found in the content of the book and that it must not be considered as an official version of the Work.

Other documents that complement these regulations and guidelines are available from the relative branches of government and the documents contained herein may be updated from time to time. It is the responsibility of the user to ensure that the information is still valid and accurate.

Author / Editor: Mindy J. Allport-Settle

Published by PharmaLogika, Inc.

Printed in the United States of America. First Printing.

ISBN 0-9830719-4-2
ISBN-13 978-0-9830719-4-5

Contents

Overview 1
    About this Book 1
        Included Documents and Features 1
            Small Group Instructor Training Course (SGITC) 1

Orientation 5
    The Role of Education in the Military 5
        Personnel Resocialization 5
    United States Army Training and Doctrine Command 6
        TRADOC History 7
        TRADOC Priorities 8
        Major Subordinate Organizations 8
    United States Army Training and Doctrine Command Analysis Center 9
        TRAC Mission Statement 9
        TRAC Organization 9
        The Discipline of Operations Research (OR) 10
        The TRAC Program 10
        Tools 11
    Operations Research 11
        Overview 12
        History 13
            Historical Origins 13
            Second World War 14
            Map of Kammhuber Line 16
            After World War II 17
        Problems Addressed with Operational Research 18
        Management Science 19
            Techniques 19
            Applications of Management Science 20

Small Group Instructor Training Course (SGITC)
Training Support Packages (Student Guide) 23
    Training Support Packages .... SG-i
    Overhead Presentation Slides and Student Handouts SG-1
        Lesson 1: Orientation L1-A1
        Lesson 2: Roles/Responsibilities/Definitions L2-A1
            Group Consensus Exercise (Problem with SGI) SG2-21
        Lesson 3: Group Development L3-A1
            Behavioral Change Chart SG3-29

Exercise B: Leadership Decision Matrix	SG3-31
Exercise C: Active Listening Exercise	SG3-33
Exercise E: Johari Window Self-Rating Sheet	SG3-35
Lesson 4: Experiential Learning Cycle	**L4-A1**
The Experiential Learning Model (Blank)	SG4-27
The Experiential Learning Model (Complete)	SG4-29
ELC Questions	SG4-31
Lesson 5: Intervention	**L5-A1**
Situation 1 - Instructor Qualifications	SG5-15
Situation 2 - War Stories and "Bull Sessions"	SG5-17
Situation 3 - Nonparticipation	SG5-19
Situation 4 - Late Student	SG5-21
Situation 5 - The "Angry Huff"	SG5-23
Situation 6 - Les Miserables	SG5-25
Situation 7 - The Filibuster	SG5-27
Situation 8 - The "Corrector"	SG5-29
Situation 9 - "I Agree"	SG5-31
Situation 10 - Critique	SG5-33
Situation 11 - Attack	SG5-35
Situation 12 - Delay	SG5-37
Situation 13 - Self-worth	SG5-39
Situation 14 - Group Norms	SG5-41
Situation 15 - Adam Ant	SG5-43
Lesson 6: Leaderless Discussions	**L6-A1**
Rules for the Generating Phase of Brainstorming	SG6a-11
Rules for Evaluating Phase of Brainstorming	SG6a-13
Performance Evaluation Checklist	SG6a-15
Performance Evaluation Checklist	SG6b-11
Performance Evaluation Checklist	SG6c-7
Lesson 7: The Conference	**L7-A1**
Performance Evaluation Checklist	SG7-7

Lesson 8: Role Playing	L8-A1
The Situation	SG8-7
Observer's Worksheet	SG8-9
Performance Evaluation Checklist	SG8-11
Lesson 9: Committee Problem Solving	L9-A1
Committee Problem Solving-Training Development Exercise	SG9-9
Performance Evaluation Checklist	SG9-11
Lesson 10: Case Studies	L10-A1
Basic Training Instructions: Group Discussion	SG10a-9
Basic Training Reading	SG10a-11
Performance Evaluation Checklist	SG10a-13
Performance Evaluation Checklist	SG10b-9
Team Instruction Sheet	SG10c-9
Performance Evaluation Checklist	SG10c-11

Overview

About this Book

The United States Army is recognized internationally as the standard for complete, efficient and effective adult education. The Army has a tradition of pioneering training systems (including computer-based training) that then transition into the corporate sector. This manual has been continuously tested and updated to successfully educate every member of the modern United States Army. The needs of the instructor, the student, and the Army are perfectly balanced. This is the model all educators should follow when developing and delivering training programs.

This book is **Volume 2** of a **2 volume set** and provides an example of how training materials should be designed and delivered for any adult education training program, but especially for any industry that is subject to government regulation.

Included Documents and Features

Small Group Instructor Training Course (SGITC)

Training Support Packages (Student Guide) .... SG-i
- Overhead Presentation Slides and Student Handouts SG-1
  - Lesson 1: Orientation L1-A1
  - Lesson 2: Roles/Responsibilities/Definitions L2-A1
    - Group Consensus Exercise (Problem with SGI) SG2-21
  - Lesson 3: Group Development L3-A1
    - Behavioral Change Chart SG3-29
    - Exercise B: Leadership Decision Matrix SG3-31
    - Exercise C: Active Listening Exercise SG3-33
    - Exercise E: Johari Window Self-Rating Sheet SG3-35

Overview

- Lesson 4: Experiential Learning Cycle L4-A1
  - The Experiential Learning Model (Blank) SG4-27
  - The Experiential Learning Model (Complete) SG4-29
  - ELC Questions SG4-31
- Lesson 5: Intervention L5-A1
  - Situation 1 - Instructor Qualifications SG5-15
  - Situation 2 - War Stories and "Bull Sessions" SG5-17
  - Situation 3 - Nonparticipation SG5-19
  - Situation 4 - Late Student SG5-21
  - Situation 5 - The "Angry Huff" SG5-23
  - Situation 6 - Les Miserables SG5-25
  - Situation 7 - The Filibuster SG5-27
  - Situation 8 - The "Corrector" SG5-29
  - Situation 9 - "I Agree" SG5-31
  - Situation 10 - Critique SG5-33
  - Situation 11 - Attack SG5-35
  - Situation 12 - Delay SG5-37
  - Situation 13 - Self-worth SG5-39
  - Situation 14 - Group Norms SG5-41
  - Situation 15 - Adam Ant SG5-43
- Lesson 6: Leaderless Discussions L6-A1
  - Rules for the Generating Phase of Brainstorming SG6a-11
  - Rules for Evaluating Phase of Brainstorming SG6a-13
  - Performance Evaluation Checklist SG6a-15
  - Performance Evaluation Checklist SG6b-11
  - Performance Evaluation Checklist SG6c-7
- Lesson 7: The Conference L7-A1
  - Performance Evaluation Checklist SG7-7
- Lesson 8: Role Playing L8-A1
  - The Situation SG8-7
  - Observer's Worksheet SG8-9

About this Book

- Performance Evaluation Checklist SG8-11
- Lesson 9: Committee Problem Solving L9-A1
  - Committee Problem Solving-Training Development Exercise
    .... SG9-9
  - Performance Evaluation Checklist SG9-11
- Lesson 10: Case Studies L10-A1
  - Basic Training Instructions: Group Discussion SG10a-9
  - Basic Training Reading SG10a-11
  - Performance Evaluation Checklist SG10a-13
  - Performance Evaluation Checklist SG10b-9
  - Team Instruction Sheet SG10c-9
  - Performance Evaluation Checklist SG10c-11

Orientation

The Role of Education in the Military

Military education and training is a process which intends to establish and improve the capabilities of military personnel in their respective roles.

Military education can be voluntary or compulsory duty. Before any person gets authorization to operate technical equipment or be on the battle field, they must take a medical and often a physical test. If passed, they may begin primary training.

The primary training is recruit training. Recruit training attempts to teach the basic information and training in techniques necessary to be an effective service member.

To achieve this, service members are drilled physically, technically and psychologically. The drill instructor has the task of making the service members fit for military use.

After finishing basic training, many service members undergo advanced training more in line with their chosen or assigned specialties. This range from navy training to studies of explosives. In advanced training, military technology and equipment is often taught.

Many large countries have several military academies, one for each branch of the service, that offer college degrees in a variety of subjects, similar to other colleges. However, academy graduates usually rank as officers, and as such have many options besides civilian work in their major subject. Higher ranking officers also have further educational opportunities.

Personnel Resocialization

While regulated industry does not seek to create soldiers and certainly does not want to employ all of the standard military training techniques, it does need to make certain personnel are trained to follow directions precisely and respect the authority of the regulatory agencies.

Resocialization is an important aspect of inducting a civilian into a military. Resocialization is a sociological concept dealing with the process of mentally and emotionally "re-training" a person so they can operate in an environment other than what they are accustomed to. Successful resocialization into a total institution involves changes to an individual's personality.

Key examples include the process of resocializing new recruits into the military so that they can operate as soldiers – or, in other words, as members of a cohesive unit. Another example is the reverse process, in which those who have become accustomed to such roles return to society after military discharge.

Resocialization from the life of a combat soldier to a civilian member of society is often difficult because of what that soldier saw and did in his military experience. In the transition from civilian to soldier, the individual is trained to solely follow the command of his superiors. In some cases commands would go against certain natural aversions (such as killing) of the individual based on one's moral and ethical principles.

A leading expert in military training methods, Grossman(2001) gives four types of training techniques used; brutalization, classical conditioning, operant conditioning and role modeling.[1] According to Grossman (2001), these techniques were meant to break down barriers to embrace a new set of norms and way of life (brutalization), condition them to pair killing with something more enjoyable and pleasurable (Classical Conditioning), repeat the stimulus-response reaction to develop a reflex (Operant Conditioning), and finally the use of a role model of a superior to provide action by example.

While leaders effectively train their soldiers to accomplish the goal of battle preparedness, these techniques increase psychological trauma experienced in veterans post-combat.[2] It is because of the evident psychological problems in post-combat situations (i.e. Post Traumatic Stress Disorder) that pose a threat to public safety because of the conditioning of the individual who might be made unstable because of his actions. Resocialization following such intense training and conditioning should be further researched and developed to better aide those discharged from the military service.

United States Army Training and Doctrine Command

Established 1 July 1973, the United States Army Training and Doctrine Command (TRADOC) is an army command of the United States Army headquartered at Fort Monroe, Virginia. It is charged with overseeing training of Army forces, the development of operational doctrine, and the development and procurement of new weapons systems. TRADOC operates 33 schools and centers at 16 Army

---

[1] Grossman, D. (2001) Trained to Kill. *Professorenforum-Journal*,2(2).
[2] Kilner, P. (2002, March). Military Leaders' Obligation to Justify Killing in War. *Military Review*, 32(2).

installations. TRADOC schools conduct 2,734 courses (81 directly in support of mobilization) and 373 language courses. The 2,734 courses include 503,164 seats for 434,424 soldiers; 34,675 other-service personnel; 7,824 international soldiers; and 26,241 civilians.[3]

TRADOC MissionThe official mission statement for TRADOC states:

> TRADOC develops the Army's Soldiers and Civilian leaders and designs, develops and integrates capabilities, concepts and doctrine in order to build a campaign-capable, expeditionary Army in support of joint warfighting capability through Army Force Generation (ARFORGEN).[4]

TRADOC is the official command component that is responsible for training and developing the United States Army.

TRADOC History

TRADOC was established as a major U.S. Army command on 1 July 1973. The new command, along with the U.S. Army Forces Command (FORSCOM), was created from the Continental Army Command (CONARC) located at Fort Monroe, VA. That action was the major innovation in the Army's post-Vietnam reorganization, in the face of realization that CONARC's obligations and span of control were too broad for efficient focus. The new organization functionally realigned the major Army commands in the continental United States. CONARC, and Headquarters, U.S. Army Combat Developments Command (CDC), situated at Fort Belvoir, VA, were discontinued, with TRADOC and FORSCOM at Fort Belvoir assuming the realigned missions. TRADOC assumed the combat developments mission from CDC, took over the individual training mission formerly the responsibility of CONARC, and assumed command from CONARC of the major Army installations in the United States housing Army training center and Army branch schools. FORSCOM assumed CONARC's operational responsibility for the command and readiness of all divisions and corps in the continental U.S. and for the installations where they were based.

Joined under TRADOC, the major Army missions of individual training and combat developments each had its own lineage. The individual training responsibility had belonged, during World War II, to Headquarters Army Ground Forces (AGF). In 1946 numbered Army areas were established in the U.S. under AGF command. At that time, the AGF moved from Washington, D.C. to Fort Monroe, VA. In March 1948, the AGR was replaced at Fort Monroe with the new

---

[3] TRADOC fact sheet available at: http://www.tradoc.army.mil/about.htm
[4] TRADOC commander on ARFORGEN, and the US Army available at: http://www.tradoc.army.mil/about.htm

Office, Chief of Army Field Forces (OCAFF). OCAFF, however, did not command the training establishment. That function was exercised by Headquarters, Department of the Army through the numbered Armies to the corps, division, and Army Training Centers. In February 1955, HQ Continental Army Command (CONARC) replaced OCAFF, assuming its missions as well as the training missions from DA. In January, HQ CONARC was redesignated U.S. Continental Army Command. Combat developments emerged as a formal Army mission in the early 1950s, and OCAFF assumed that role in 1952. In 1955, CONARC assumed the mission. In 1962, HQ U.S. Army Combat Development Command (CDC) was established to bring the combat developments function under one major Army command.[5]

TRADOC Priorities

1. Leader Development
2. Initial Military Training
3. Concepts and Capabilities Integration
4. Human Capital Enterprise
5. Army Training and Learning Concept
6. Doctrine

Major Subordinate Organizations

- U.S. Army Accessions Command (USAAC)
- U.S. Army Capabilities Integration Center (ARCIC)
- U.S. Army Combined Arms Center (CAC)
- U.S. Army Combined Arms Support Command (CASCOM)
- U.S. Army TRADOC Analysis Center (TRAC)
- U.S. Army Center for Army Lessons Learned (CALL)
- U.S. Army Initial Military Training (IMT)

---

[5] TRADOC history available at: http://www.tradoc.army.mil/about.htm

United States Army Training and Doctrine Command Analysis Center

The United States Army Training and Doctrine Command Analysis Center (TRAC) is an analysis agency of the United States Army. TRAC conducts research on potential military operations worldwide to inform decisions about the most challenging issues facing the Army and the Department of Defense (DoD). TRAC relies upon the intellectual capital of a highly skilled workforce of military and civilian personnel to execute its mission.

TRAC conducts operations research (OR) on a wide range of military topics, some contemporary but most often set 5 to 15 years in the future. How should Army units be organized? What new systems should be procured? How should soldiers and commanders be trained? What are the costs and benefits of competing options? What are the potential risks and rewards of a planned military course of action? TRAC directly supports the mission of the Army's Training and Doctrine Command (TRADOC), to develop future concepts and requirements while also serving the decision needs of many military clients.

TRAC Mission Statement

To produce relevant and credible operations analysis to inform decisions. TRAC serves many clients and has many stakeholders, but has only one shareholder: The American Soldier.

TRAC Organization

TRAC is led by a civilian SES director, subordinate to the Commanding General of the US Army Training and Doctrine Command, and comprises four centers:[6]

- TRAC-Fort Leavenworth (TRAC-FLVN), led by a civilian SES director, is co-located with TRAC headquarters at Fort Leavenworth, Kansas (i.e. US Army Combined Arms Center (CAC)) and has traditionally conducted analysis at the operational level (i.e. Corps and Division).

- TRAC-White Sands Missile Range (TRAC-WSMR), led by a civilian SES director, is located at White Sands Missile Range in New Mexico and has traditionally conducted analysis at the tactical level (i.e. Brigade and below).

- TRAC-Lee, led by a civilian GS-15, is located at Fort Lee, VA and has traditionally conducted analysis in the area of Logistics.

---

[6] TRADOC Regulation 10-5-7, Organization and Functions U.S. Army TRADOC Analysis Center, 29 December 2005

Orientation

- TRAC-Monterey, led by a Lieutenant Colonel (LTC), is co-located with the Naval Postgraduate School (NPS) in Monterey, CA and has traditionally utilized the resources of NPS to conduct research into new models and methodologies.

Each center director is subordinate to the TRAC director.

TRAC headquarters has three components:

- The director's staff element comprising the deputy director, an O-6 colonel, and administrative assistants
- The Program and Resources Directorate (PRD), in coordination with center directors, oversees the day-to-day operations of all TRAC elements
- The Methods and Research Office (MRO) conducts research into analysis of cutting edge military operations.

The Discipline of Operations Research (OR)

The discipline of Operations Research (OR) is built upon the collaboration of interdisciplinary team members who have mutually supporting knowledge, skills and experiences pertinent to the study problem. The TRAC building blocks of military operations analysis are future scenarios; leading edge models and simulations; realistic data about systems, forces, and behavior; and skilled operations analysts. Leading a core team of analysts, a TRAC Study Director may receive support from other TRADOC and Army agencies, and from other government agencies and industry as well. TRAC adheres to the proven principles of scientific inquiry and applies the problem solving model to perform its analysis.

The TRAC Program

The TRAC program of operations research and analysis is forward-looking and addresses a wide range of military topics. The analysis is conducted within a joint framework of combined arms operations across a full spectrum of missions and environments. TRAC leads TRADOC's major studies of new warfighting operations and organization (O&O) concepts and requirements. TRAC leads the Army's analysis of Advanced Warfighting Experiments (AWEs), and the Army's Analysis of Alternatives (AoA). The analysis topics span doctrine, training, leader development, organization, materiel, and soldier support.

Scenarios are used by the U.S. Army for education, training and force development. Director, TRAC is the TRADOC executive agent for development of scenarios for use in studies and analysis.[7] TRAC develops scenarios of potential military operations set in the future for use in modeling and analysis. TRAC relies upon input and assistance from many Army and DoD agencies, other Services and the Combatant Commanders to develop and apply a family of scenarios depicting joint operations of corps and divisions, and brigades and battalions. The family of scenarios undergoes continual review and change in anticipation of emerging threats and new operational environments around the world based on intelligence estimates.

Tools

Military operations are highly complex processes and typically must be modeled in order to be analyzed. The analytic tools may take the form of:

- table-top map games,
- human-in-the-loop (HITL),
- (HITL) simulations and simulators,
- closed-form Models & Simulations (M&S), and
- controlled field experiments.

TRAC develops and maintains a class of warfighting M&S referred to as force-on-force, ranging from individual objects (e.g., soldier, weapon, terrain feature) to aggregated objects (e.g., battalions) at corps level. TRAC M&S represent the Army's de facto standards for force-on-force M&S and are widely used by military, industry and allies. TRAC is a significant contributor to advanced M&S research and improved modeling methodologies in the military.

Operations Research

Operations research (also referred to as decision science, or management science) is an interdisciplinary mathematical science that focuses on the effective use of technology by organizations. In contrast, many other science & engineering disciplines focus on technology giving secondary considerations to its use.

Employing techniques from other mathematical sciences — such as mathematical modeling, statistical analysis, and mathematical optimization — operations research arrives at optimal or near-optimal solutions to complex decision-making problems. Because of its emphasis on human-technology interaction and because of its focus on practical applications, operations research has overlap with other

---

[7] TRADOC regulation TR 71-4. United States Army Training And Doctrine Command Standard Scenarios For Capability Developments, 23 September 2008.

Orientation

disciplines, notably industrial engineering and management science, and draws on psychology and organization science. Operations Research is often concerned with determining the maximum (of profit, performance, or yield) or minimum (of loss, risk, or cost) of some real-world objective. Originating in military efforts before World War II, its techniques have grown to concern problems in a variety of industries.[8]

Overview

Operational research encompasses a wide range of problem-solving techniques and methods applied in the pursuit of improved decision-making and efficiency.[9] Some of the tools used by operational researchers are statistics, optimization, probability theory, queuing theory, game theory, graph theory, decision analysis, mathematical modeling and simulation. Because of the computational nature of these fields, OR also has strong ties to computer science and analytics. Operational researchers faced with a new problem must determine which of these techniques are most appropriate given the nature of the system, the goals for improvement, and constraints on time and computing power.

Work in operational research and management science may be characterized as one of three categories:[10]

- Fundamental or foundational work takes place in three mathematical disciplines: probability, optimization, and dynamical systems theory.
- Modeling work is concerned with the construction of models, analyzing them mathematically, implementing them on computers, solving them using software tools, and assessing their effectiveness with data. This level is mainly instrumental, and driven mainly by statistics and econometrics.
- Application work in operational research, like other engineering and economics' disciplines, attempts to use models to make a practical impact on real-world problems.

The major subdisciplines in modern operational research, as identified by the journal Operations Research,[11] are:

- Computing and information technologies
- Decision analysis

---

[8] Available at: http://www.hsor.org/what_is_or.cfm
[9] Available at: http://www.bls.gov/oco/ocos044.htm
[10] "What is Management Science Research?" University of Cambridge 2008. Retrieved 5 June 2008.
[11] Available at: http://www3.informs.org/site/OperationsResearch/index.php?c=10&kat=Forthcoming+Papers

- Environment, energy, and natural resources
- Financial engineering
- Manufacturing, service sciences, and supply chain management
- Marketing Engineering [12]
- Policy modeling and public sector work
- Revenue management
- Simulation
- Stochastic models
- Transportation

History

As a formal discipline, operational research originated in the efforts of military planners during World War II. In the decades after the war, the techniques began to be applied more widely to problems in business, industry and society. Since that time, operational research has expanded into a field widely used in industries ranging from petrochemicals to airlines, finance, logistics, and government, moving to a focus on the development of mathematical models that can be used to analyze and optimize complex systems, and has become an area of active academic and industrial research.[13]

Historical Origins

In the World War II era, operational research was defined as "a scientific method of providing executive departments with a quantitative basis for decisions regarding the operations under their control."[14] Other names for it included

---

[12] Available at: http://www.decisionpro.biz
[13] Available at: http://www.hsor.org/what_is_or.cfm
[14] "Operational Research in the British Army 1939–1945, October 1947, Report C67/3/4/48, UK National Archives file WO291/1301 Quoted on the dust-jacket of: Morse, Philip M, and Kimball, George E, Methods of Operations Research, 1st Edition Revised, pub MIT Press & J Wiley, 5th printing, 1954.

Orientation

operational analysis (UK Ministry of Defence from 1962)[15] and quantitative management.[16]

Prior to the formal start of the field, early work in operational research was carried out by individuals such as Charles Babbage. His research into the cost of transportation and sorting of mail led to England's universal "Penny Post" in 1840, and studies into the dynamical behaviour of railway vehicles in defence of the GWR's broad gauge.[17] Percy Bridgman brought operational research to bear on problems in physics in the 1920s and would later attempt to extend these to the social sciences.[18] The modern field of operational research arose during World War II.

Modern operational research originated at the Bawdsey Research Station in the UK in 1937 and was the result of an initiative of the station's superintendent, A. P. Rowe. Rowe conceived the idea as a means to analyse and improve the working of the UK's early warning radar system, Chain Home (CH). Initially, he analyzed the operating of the radar equipment and its communication networks, expanding later to include the operating personnel's behaviour. This revealed unappreciated limitations of the CH network and allowed remedial action to be taken.[19]

Scientists in the United Kingdom including Patrick Blackett later Lord Blackett OM PRS, Cecil Gordon, C. H. Waddington, Owen Wansbrough-Jones, Frank Yates, Jacob Bronowski and Freeman Dyson, and in the United States with George Dantzig looked for ways to make better decisions in such areas as logistics and training schedules. After the war it began to be applied to similar problems in industry.

Second World War

Patrick BlackettDuring the Second World War close to 1,000 men and women in Britain were engaged in operational research. About 200 operational research scientists worked for the British Army.[20]

---

[15] UK National Archives Catalogue for WO291 lists a War Office organisation called Army Operational Research Group (AORG) that existed from 1946 to 1962. "In January 1962 the name was changed to Army Operational Research Establishment (AORE). Following the creation of a unified Ministry of Defence, a tri-service operational research organisation was established: the Defence Operational Research Establishment (DOAE) which was formed in 1965, and it absorbed the Army Operational Research Establishment based at West Byfleet."
[16] Available at: http://brochure.unisa.ac.za/myunisa/data/subjects/Quantitative%20Management.pdf
[17] M.S. Sodhi, "What about the 'O' in O.R.?" OR/MS Today, December, 2007, p. 12, http://www.lionhrtpub.com/orms/orms-12-07/frqed.html
[18] P. W. Bridgman, The Logic of Modern Physics, The MacMillan Company, New York, 1927
[19] Available at: http://www.britannica.com/EBchecked/topic/682073/operations-research/68171/History#ref22348
[20] Kirby, p. 117

Patrick Blackett worked for several different organizations during the war. Early in the war while working for the Royal Aircraft Establishment (RAE) he set up a team known as the "Circus" which helped to reduce the number of anti-aircraft artillery rounds needed to shoot down an enemy aircraft from an average of over 20,000 at the start of the Battle of Britain to 4,000 in 1941.[21]

In 1941 Blackett moved from the RAE to the Navy, first to the Royal Navy's Coastal Command, in 1941 and then early in 1942 to the Admiralty.[22] Blackett's team at Coastal Command's Operational Research Section (CC-ORS) included two future Nobel prize winners and many other people who went on to be preeminent in their fields.[23] They undertook a number of crucial analyses that aided the war effort. Britain introduced the convoy system to reduce shipping losses, but while the principle of using warships to accompany merchant ships was generally accepted, it was unclear whether it was better for convoys to be small or large. Convoys travel at the speed of the slowest member, so small convoys can travel faster. It was also argued that small convoys would be harder for German U-boats to detect. On the other hand, large convoys could deploy more warships against an attacker. Blackett's staff showed that the losses suffered by convoys depended largely on the number of escort vessels present, rather than on the overall size of the convoy. Their conclusion, therefore, was that a few large convoys are more defensible than many small ones.[24]

While performing an analysis of the methods used by RAF Coastal Command to hunt and destroy submarines, one of the analysts asked what colour the aircraft were. As most of them were from Bomber Command they were painted black for nighttime operations. At the suggestion of CC-ORS a test was run to see if that was the best colour to camouflage the aircraft for daytime operations in the grey North Atlantic skies. Tests showed that aircraft painted white were on average not spotted until they were 20% closer than those painted black. This change indicated that 30% more submarines would be attacked and sunk for the same number of sightings.[25]

Other work by the CC-ORS indicated that on average if the trigger depth of aerial delivered depth charges (DCs) was changed from 100 feet to 25 feet, the kill ratios would go up. The reason was that if a U-boat saw an aircraft only shortly before it arrived over the target then at 100 feet the charges would do no damage (because the U-boat wouldn't have time to descend as far as 100 feet), and if it saw the aircraft a long way from the target it had time to alter course under water so the

---

[21] Kirby, pp. 91–94
[22] Kirby, p. 96,109
[23] Kirby, p. 96
[24] "Numbers are Essential": Victory in the North Atlantic Reconsidered, March–May 1943
[25] Kirby, p. 101

chances of it being within the 20 feet kill zone of the charges was small. It was more efficient to attack those submarines close to the surface when these targets' locations were better known than to attempt their destruction at greater depths when their positions could only be guessed. Before the change of settings from 100 feet to 25 feet, 1% of submerged U-boats were sunk and 14% damaged. After the change, 7% were sunk and 11% damaged. (If submarines were caught on the surface, even if attacked shortly after submerging, the numbers rose to 11% sunk and 15% damaged). Blackett observed "there can be few cases where such a great operational gain had been obtained by such a small and simple change of tactics."[26]

Bomber Command's Operational Research Section (BC-ORS), analysed a report of a survey carried out by RAF Bomber Command.[citation needed] For the survey, Bomber Command inspected all bombers returning from bombing raids over Germany over a particular period. All damage inflicted by German air defences was noted and the recommendation was given that armour be added in the most heavily damaged areas. Their suggestion to remove some of the crew so that an aircraft loss would result in fewer personnel loss was rejected by RAF command. Blackett's team instead made the surprising and counter-intuitive recommendation that the armour be placed in the areas which were completely untouched by damage in the bombers which returned. They reasoned that the survey was biased, since it only included aircraft that returned to Britain. The untouched areas of returning aircraft were probably vital areas, which, if hit, would result in the loss of the aircraft.[27]

Map of Kammhuber Line

When Germany organized its air defences into the Kammhuber Line, it was realised that if the RAF bombers were to fly in a bomber stream they could overwhelm the night fighters who flew in individual cells directed to their targets by ground controllers. It was then a matter of calculating the statistical loss from collisions against the statistical loss from night fighters to calculate how close the bombers should fly to minimize RAF losses.[28]

The "exchange rate" ratio of output to input was a characteristic feature of operational research. By comparing the number of flying hours put in by Allied aircraft to the number of U-boat sightings in a given area, it was possible to redistribute aircraft to more productive patrol areas. Comparison of exchange rates established "effectiveness ratios" useful in planning. The ratio of 60 mines laid per ship sunk was common to several campaigns: German mines in British

---

[26] Kirby, pp. 102,103
[27] "Dirty Little Secrets of the Twentieth Century" authored by James F. Dunnigan on pages 215-217
[28] Available at: http://www.raf.mod.uk/bombercommand/thousands.html

ports, British mines on German routes, and United States mines in Japanese routes.[29]

Operational research doubled the on-target bomb rate of B-29s bombing Japan from the Marianas Islands by increasing the training ratio from 4 to 10 percent of flying hours; revealed that wolf-packs of three United States submarines were the most effective number to enable all members of the pack to engage targets discovered on their individual patrol stations; revealed that glossy enamel paint was more effective camouflage for night fighters than traditional dull camouflage paint finish, and the smooth paint finish increased airspeed by reducing skin friction.[30]

On land, the operational research sections of the Army Operational Research Group (AORG) of the Ministry of Supply (MoS) were landed in Normandy in 1944, and they followed British forces in the advance across Europe. They analysed, among other topics, the effectiveness of artillery, aerial bombing, and anti-tank shooting.

After World War II

With expanded techniques and growing awareness of the field at the close of the war, operational research was no longer limited to only operational, but was extended to encompass equipment procurement, training, logistics and infrastructure.

Academic Denis Bouyssou describes the historical development of operational research from the 1940s to the 1970s as follows. "The historical development of Operational Research (OR) is traditionally seen as the succession of several phases: the 'heroic times' of the Second World War, the 'Golden Age' between the fifties and the sixties during which major theoretical achievements were accompanied by a widespread diffusion of OR techniques in private and public organisations, a 'crisis' followed by a 'decline' starting with the late sixties, a phase during which OR groups in firms progressively disappeared while academia became less and less concerned with the applicability of the techniques developed".[31]

Individuals such as Stafford Beer and George Dantzig pioneered early academic efforts in operational research.

---

[29] Milkman, Raymond H. (May 1968). Operations Research in World War II. United States Naval Institute Proceedings.
[30] Milkman, Raymond H. (May 1968). Operations Research in World War II. United States Naval Institute Proceedings.
[31] Bouyssou, Denis, "Questioning the history of operational research in order to prepare its future" available at: http://hal.ccsd.cnrs.fr/docs/00/02/86/41/PDF/cahierLamsade196.pdf

Orientation

Problems Addressed with Operational Research

- Critical path analysis or project planning: identifying those processes in a complex project which affect the overall duration of the project
- Floorplanning: designing the layout of equipment in a factory or components on a computer chip to reduce manufacturing time (therefore reducing cost)
- Network optimization: for instance, setup of telecommunications networks to maintain quality of service during outages
- Allocation problems
- Bayesian search theory: looking for a target
- Optimal search
- Routing, such as determining the routes of buses so that as few buses are needed as possible
- Supply chain management: managing the flow of raw materials and products based on uncertain demand for the finished products
- Efficient messaging and customer response tactics
- Automation: automating or integrating robotic systems in human-driven operations processes
- Globalization: globalizing operations processes in order to take advantage of cheaper materials, labor, land or other productivity inputs
- Transportation: managing freight transportation and delivery systems (Examples: LTL Shipping, intermodal freight transport)
- Scheduling:
    - personnel staffing
    - manufacturing steps
    - project tasks
    - network data traffic: these are known as queueing models or queueing systems.
    - sports events and their television coverage
    - blending of raw materials in oil refineries
    - determining optimal prices, in many retail and B2B settings, within the disciplines of pricing science

Operational research is also used extensively in government where evidence-based policy is used.

Management Science

In 1967 Stafford Beer characterized the field of management science as "the business use of operations research".[32] However, in modern times the term management science may also be used to refer to the separate fields of organizational studies or corporate strategy. Like operational research itself, management science (MS), is an interdisciplinary branch of applied mathematics devoted to optimal decision planning, with strong links with economics, business, engineering, and other sciences. It uses various scientific research-based principles, strategies, and analytical methods including mathematical modeling, statistics and numerical algorithms to improve an organization's ability to enact rational and meaningful management decisions by arriving at optimal or near optimal solutions to complex decision problems. In short, management sciences help businesses to achieve their goals using the scientific methods of operational research.

The management scientist's mandate is to use rational, systematic, science-based techniques to inform and improve decisions of all kinds. Of course, the techniques of management science are not restricted to business applications but may be applied to military, medical, public administration, charitable groups, political groups or community groups.

Management science is concerned with developing and applying models and concepts that may prove useful in helping to illuminate management issues and solve managerial problems, as well as designing and developing new and better models of organizational excellence.[33]

The application of these models within the corporate sector became known as Management science.[34]

Techniques

Some of the fields that have considerable overlap with Management Science include:

- Data mining

---

[32] Stafford Beer (1967) Management Science: The Business Use of Operations Research
[33] "What is Management Science?" Lancaster University, 2008. Available at: http://www.lums.lancs.ac.uk/departments/mansci/DeptProfile/WhatisManSci/.
[34] "What is Management Science?" The University of Tennessee, 2006. Available at: http://bus.utk.edu/soms/information/whatis_msci.html.

Orientation

- Decision analysis
- Engineering
- Forecasting
- Game theory
- Industrial engineering
- Logistics
- Mathematical modeling
- Optimization
- Probability and statistics
- Project management
- Simulation
- Social network/Transportation forecasting models
- Supply chain management
- Financial engineering

Applications of Management Science

Applications of management science are abundant in industry as airlines, manufacturing companies, service organizations, military branches, and in government. The range of problems and issues to which management science has contributed insights and solutions is vast. It includes:[35]

- Scheduling airlines, including both planes and crew,
- Deciding the appropriate place to site new facilities such as a warehouse, factory or fire station,
- Managing the flow of water from reservoirs,
- Identifying possible future development paths for parts of the telecommunications industry,
- Establishing the information needs and appropriate systems to supply them within the health service, and

---

[35] "What is Management Science?" Lancaster University, 2008. Available at: http://www.lums.lancs.ac.uk/departments/mansci/DeptProfile/WhatisManSci/.

- Identifying and understanding the strategies adopted by companies for their information systems

Management science is also concerned with so-called "soft-operational analysis," which concerns methods for strategic planning, strategic decision support, and Problem Structuring Methods (PSM). In dealing with these sorts of challenges mathematical modeling and simulation are not appropriate or will not suffice. Therefore, during the past 30 years, a number of non-quantified modeling methods have been developed. These include:

- Stakeholder based approaches including metagame analysis and drama theory
- Morphological analysis and various forms of influence diagrams.
- Approaches using cognitive mapping
- The Strategic Choice Approach
- Robustness analysis

Small Group Instructor Training Course (SGITC)

Training Support Packages (Student Guide)

SMALL GROUP INSTRUCTOR TRAINING COURSE
(SGITC)

JUNE 1998

This page intentionally left blank.

SMALL GROUP INSTRUCTOR TRAINING COURSE
(SGITC)

TRAINING SUPPORT PACKAGES

This page intentionally left blank.

SMALL GROUP INSTRUCTOR TRAINING COURSE

(SGITC)

TRAINING SUPPORT PACKAGE FOR

LESSON 1:

ORIENTATION

This page intentionally left blank.

LESSON 1
ORIENTATION

Small Group Instruction

Small Group Instruction

TERMINAL OBJECTIVE

Identify course content and course requirements.

L1-A2

Small Group Instruction

ENABLING OBJECTIVE 1

Communicate from your TNET site to the host site.

L1-A3

VTT SITE

Small Group Instruction

TERMINAL LEARNING OBJECTIVE

L1-A4

Small Group Instruction

ENABLING OBJECTIVE 2

Describe the instructional blocks of the SGI course.

Small Group Instruction

FOUR INSTRUCTIONAL BLOCKS

- Group and Individual Dynamics
- SGI Methods
- Roles, Responsibilities, and Definitions
- Experiential Learning Cycle (ELC)

Small Group Instruction

ROLES, RESPONSIBILITIES, AND DEFINITIONS

- Definitions
- Roles and responsibilities
- Learning theory application

Small Group Instruction

GROUP AND INDIVIDUAL DYNAMICS

- Three stages of group development
- Nonproductive behaviors and intervention strategies
- Self-awareness tools

Small Group Instruction

EXPERIENTIAL LEARNING CYCLE (ELC)

- Process-Content model
- Five phases of adult learning

Small Group Instruction

SGI METHODS

- Leaderless discussions
- Conference
- Role playing
- Committee problem solving
- Case discussions

Small Group Instruction

ENABLING OBJECTIVE 3

Identify the course requirements.

L1-A11

Small Group Instruction

COURSE REQUIREMENTS

- Apply SGI information
- Participate in small groups
- Present an SGI method lesson
- Adapt/deliver lesson to group
- Evaluate peers and course

Small Group Instruction

EVALUATION

- Group performance
- Performance Evaluation Checklist
- One-on-one counseling
- End-of-course evaluation

Small Group Instruction

INTRODUCTIONS

- Job/duty position
- Family status
- How to address
- Special likes
- SGI experience

SMALL GROUP INSTRUCTOR TRAINING COURSE

(SGITC)

TRAINING SUPPORT PACKAGE FOR

LESSON 2:

ROLES/RESPONSIBILITIES/DEFINITIONS

This page intentionally left blank.

Small Group Instruction

LESSON 2
ROLES, RESPONSIBILITIES, AND DEFINITIONS

Small Group Instruction

TERMINAL OBJECTIVE

Identify SGI roles, responsibilities, and definitions.

L2-A2

Small-Group Instruction

ENABLING OBJECTIVE 1

Define Small Group Instruction.

L2-A3

Small Group Instruction

WHAT IS IT?

L2-A4

Small Group Instruction

ENABLING OBJECTIVE 2

Describe an SGL and the three roles of the SGL.

L2-A5

Small Group Instruction

SME ROLE

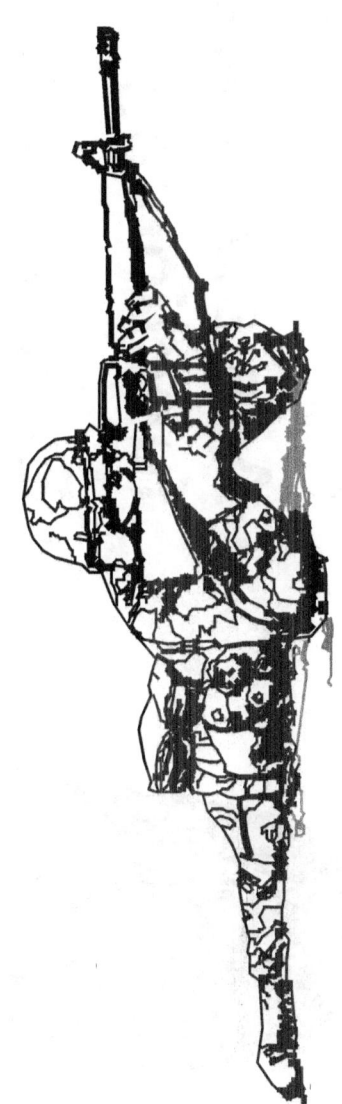

- Prepares material
- Selects group method
- Influences content

Small Group Instruction

FACILITATOR ROLE

- Stimulates group interaction
- Assists group functioning
- Expands group participation and development

L2-A8

Small Group Instruction

OBSERVER ROLE

- Observes group interaction
- Serves as base for involvement as SME or facilitator

Small Group Instruction

ENABLING OBJECTIVE 3

Identify the roles and responsibilities of the small group members.

Small Group Instruction

GROUP MEMBERS RESPONSIBLE FOR...

- Group success
- Individual success
- Individual and group participation

Small Group Instruction

GROUP ROLES

Group member

Recorder

Student Discussion Leader (SDL)

Small Group Instruction

SDL ROLE

- Assigned lesson by SGL
- Delivers lessons
- Keeps group focused

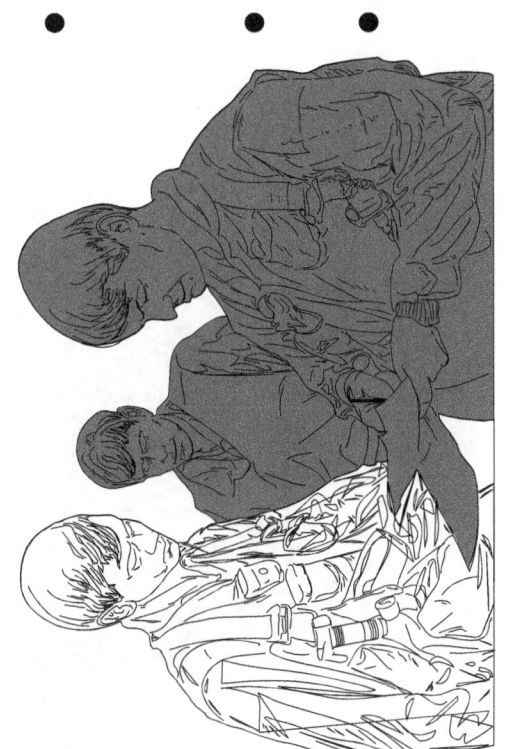

Small Group Instruction

RECORDER ROLE

- Documents group proceedings
- Verifies accuracy of notes

Small Group Instruction

GROUP MEMBER ROLE

- Participates in discussions
- Expresses facts, feelings, and opinions
- Focuses on mission accomplishment

Small Group Instruction

ENABLING OBJECTIVE 4

Apply SGI learning theory to your roles and responsibilities.

L2-A16

Small Group Instruction

CONDITIONS FOR LEARNING

- Climate
- Controlled observation
- Realistic situations
- Opportunity for experimentation
- Objective performance analysis

L2-A17

Small Group Instruction

LESSON REVIEW

Identify SGI roles, responsibilities, and definitions.

L2-A18

LESSON 2

Student Handouts and Exercises

Handout   **Page**

Group Consensus Exercise (Problem with SGI)..SG2-21

NOT USED

SGITC Lesson 2 - Roles/Responsibilities/Definitions

Group Consensus Exercise (Problem with SGI)

Instructions

a. Break into small groups (4 to 6). Conduct an abbreviated Topic Discussion on what is the biggest problem with small group instruction.

b. Discuss the topic for approximately 10 minutes, then record your top/number 1 problem with SGI.

c. Ensure the group comes to consensus.

NOT USED

SMALL GROUP INSTRUCTOR TRAINING COURSE

(SGITC)

TRAINING SUPPORT PACKAGE FOR

LESSON 3:

GROUP DEVELOPMENT

This page intentionally left blank.

Small Group Instruction

LESSON 3
GROUP DEVELOPMENT

L3-A1

Small Group Instruction

TERMINAL OBJECTIVES

Describe how small groups develop.

Describe how leadership and learning conditions affect group synergy and productivity.

Small Group Instruction

ENABLING OBJECTIVE 1

Describe the stages of group development.

Small Group Instruction

DEPENDENT/INCLUSION/ ACCEPTANCE STAGE

- Polite
- Seek instructor guidance
- Seek instructor approval

L3-A4

Small Group Instruction

INDEPENDENT/CONTROL/ INFLUENCE STAGE

- Argumentative
- Challenges SGL
- Forces group decisions

Small Group Instruction

INTERDEPENDENT/ COHESION/AFFECTION STAGE

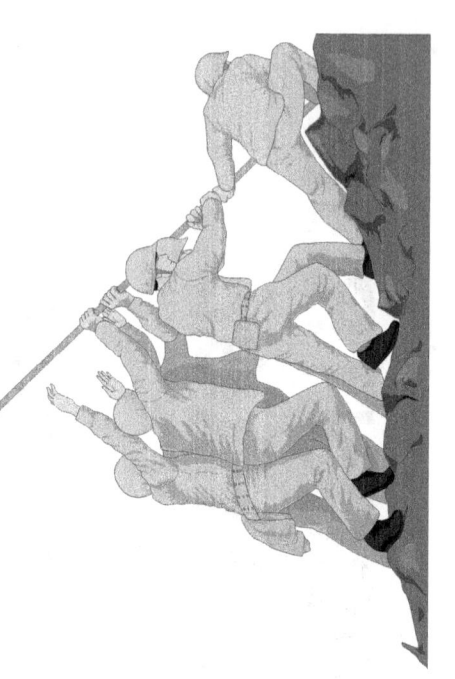

- Highly productive
- Cooperative
- Solves problems

Small Group Instruction

DEVELOPMENT PRINCIPLES

- Groups vary
- SGL adheres to roles
- No short cuts

Small Group Instruction

GROUP BEHAVIORAL DIMENSIONS

Norms

Decision making

Feedback

Structure

Influence

Competition

L3-A8

Small Group Instruction

ENABLING OBJECTIVE 2

Identify leadership styles and the effects of selected leader actions differentiating each style.

L3-A9

Small Group Instruction

PRACTICAL EXERCISE

L3-A10

Small Group Instruction

LEADERSHIP STYLES

- **Directing** = No subordinate input.

- **Participating** = Opinions from subordinates, but you make decisions.

- **Delegating** = Subordinates make most decisions.

Small Group Instruction

CONDITIONS FOR LEARNING

- Climate
- Controlled observation
- Realistic situations
- Opportunity for experimentation
- Objective performance analysis

Small Group Instruction

GROUP FORCES

- Group goals
- Group cohesiveness
- Group norms
- Communication system

Small Group Instruction

GROUP FUNCTIONS

- Resource integration
- Social motivation
- Social influence

Small Group Instruction

SUGGESTED RULES

❓ ❓ ❓ ❓ ❓

L3-A15

Small Group Instruction

ENABLING OBJECTIVE 3

Demonstrate active listening skills.

L3-A16

Small Group Instruction

ACTIVE LISTENING SKILLS

- Paraphrasing
- Checking perceptions
- Withholding evaluation

Small Group Instruction

ENABLING OBJECTIVE 4

Identify how perceptions interfere with honest communication in the small group.

L3-A18

Small Group Instruction

THE JOHARI WINDOW

	I know	I don't know
They know	Arena	Blind Spot
They don't know	Facade	Unknown

Unconscious

GROUP

L3-A19

Small Group Instruction

THE JOHARI WINDOW

	I know	I don't know
GROUP They know	Arena	Blind Spot
They don't know	Facade	Unknown

[Unconscious]

Small Group Instruction

THE JOHARI WINDOW

	I know	I don't know
GROUP They know	Arena	Blind Spot
They don't know	Facade	Unknown

Unconscious

L3-A21

Small Group Instruction

THE JOHARI WINDOW

	I know	I don't know
They know	Arena	Blind Spot
They don't know	Facade	Unknown

[Unconscious]

GROUP

Small Group Instruction

THE JOHARI WINDOW

	I know	I don't know
G R O U P They know	Arena	Blind Spot
They don't know	Facade	Unknown *(Unconscious)*

Small Group Instruction

REVIEW

Stages of Small Group Development:

- **Dependent/Inclusion/Acceptance**
  (Polite and seeks guidance/approval.)

- **Independent/Control/Influence**
  (Argues, challenges, and forces decisions.)

- **Interdependent/Cohesion/Affection**
  (Cooperative and productive.)

Small Group Instruction

REVIEW

Leadership Styles

- **Directing** = No subordinate input.

- **Participating** = Opinions from subordinates, but you make decisions.

- **Delegating** = Subordinates make most decisions.

L3-A25

Small Group Instruction

JOHARI WINDOW REVIEW

- Measures honesty and disclosure.
- Panes change size as trust and commitment increase.
- Decreasing the "Facade" increases the "Arena" to foster openness.

LESSON 3

Student Handouts and Exercises

Handout	Page
Behavioral Change Chart	SG3-29
Exercise B: Leadership Decision Matrix	SG3-31
Exercise C: Active Listening Exercise	SG3-33
Exercise E: Johari Window Self-Rating Sheet	SG3-35

NOT USED

Behavioral Change Chart

Use of this chart Use this chart to determine the stage of group development by observing the behaviors of the group.

Chart This chart compares six group behaviors with the stage of group development.

Behavior	D/I/A Stage	I/C/I Stage	I/C/A Stage
Norms	Norms are developed.	Norms are broken.	Norms are examined as a group.
Structure	Group looks to instructor.	Imposed on group by student or students.	Group looks to themselves.
Decision Making	Autocratic and minority.	Autocratic, minority, and majority.	Consensus.
Influence	Covert, reference to authority.	Overt, arguments.	Shared, goes to student for task at hand.
Feedback	Little or none.	Some, but doesn't conform to rules.	Give and receive. Conforms to rules.
Competition	Can't win. Avoidance attitude.	Must win. Belligerent attitude.	All win. Cooperative attitude.

NOT USED

SGITC   Lesson 3 - Group Development

Exercise B: Leadership Decision Matrix Handout

Instructions   Select the style of leadership to use in the following situations in the small group environment. Select only one answer. Come to consensus and report back in 15 minutes. Be prepared to discuss your answers.

Check one answer only

Situation	Dir	Par	Del
1. Morning of the 1st day, group asks questions.			
2. Conference style of SGI delivery.			
3. Midway through class, group becomes unruly.			
4. New student joins class after 3d day of 2 week class.			
5. Mature group off track on task completion.			
6. AAR or ELC after lesson delivery.			
7. One student does not participate during several group exercises.			
8. Afternoon of last day of class.			
9. Group performs hazardous task.			

NOT USED

Exercise C: Active Listening

Step 1	A list of controversial topics follows. Assign one per group member. Abortion, Bosnia, Concealed Weapon Law, Buying American, Gay Teachers, Smoke Breaks, Standing for the National Anthem, Baseball Strike, Football Teams Moving.
Step 2	Each person in your group will have a turn being the speaker. The speaker will talk for 3 minutes, selecting a topic from the list you prepared or choosing one of his/her own.

Continued on next page

Lesson 3 Group Development   SGITC

Exercise C: Active Listening Exercise, Continued

Risk of communicating nonacceptance

The communication of mutual acceptance is vital to developing and maintaining work and personal relationships. However, various ways of responding to situations run the risk of communicating nonacceptance. To understand another person's point of view effectively, you must show your openness to that communication. According to author Gordon, most people, during a listening situation, commonly respond in one or more of the following twelve ways:

1. Ordering, Directing: "You have to . . . "
2. Warning, Threatening: "You better not . . ."
3. Preaching, Moralizing: "You ought to . . ."
4. Advising, Giving Solutions: "Why don't you . . ."
5. Lecturing, Informing: "Here are the facts . . ."
6. Evaluating, Blaming: "You're wrong . . ."
7. Praising, Agreeing: "You're right . . ."
8. Name-calling, Shaming: "You're stupid . . ."
9. Interpreting, Analyzing: "What you need . . ."
10. Sympathizing, Supporting: "You'll be OK. . ."
11. Questioning, Probing: "Why did you . . ."
12. Withdrawing, Avoiding: "Let's forget it . . ."

SGITC   Lesson 3 - Group Development

Exercise E: Johari Window Self-Rating Sheet

Instructions Recall the last subgroup exercise and think of how much you solicited feedback from the group. Record this rating vertically on the form. Think about how much feedback about yourself (disclosure) you gave to the group. Record this rating horizontally on the form. The results are your windows. Share these rating with your group members.

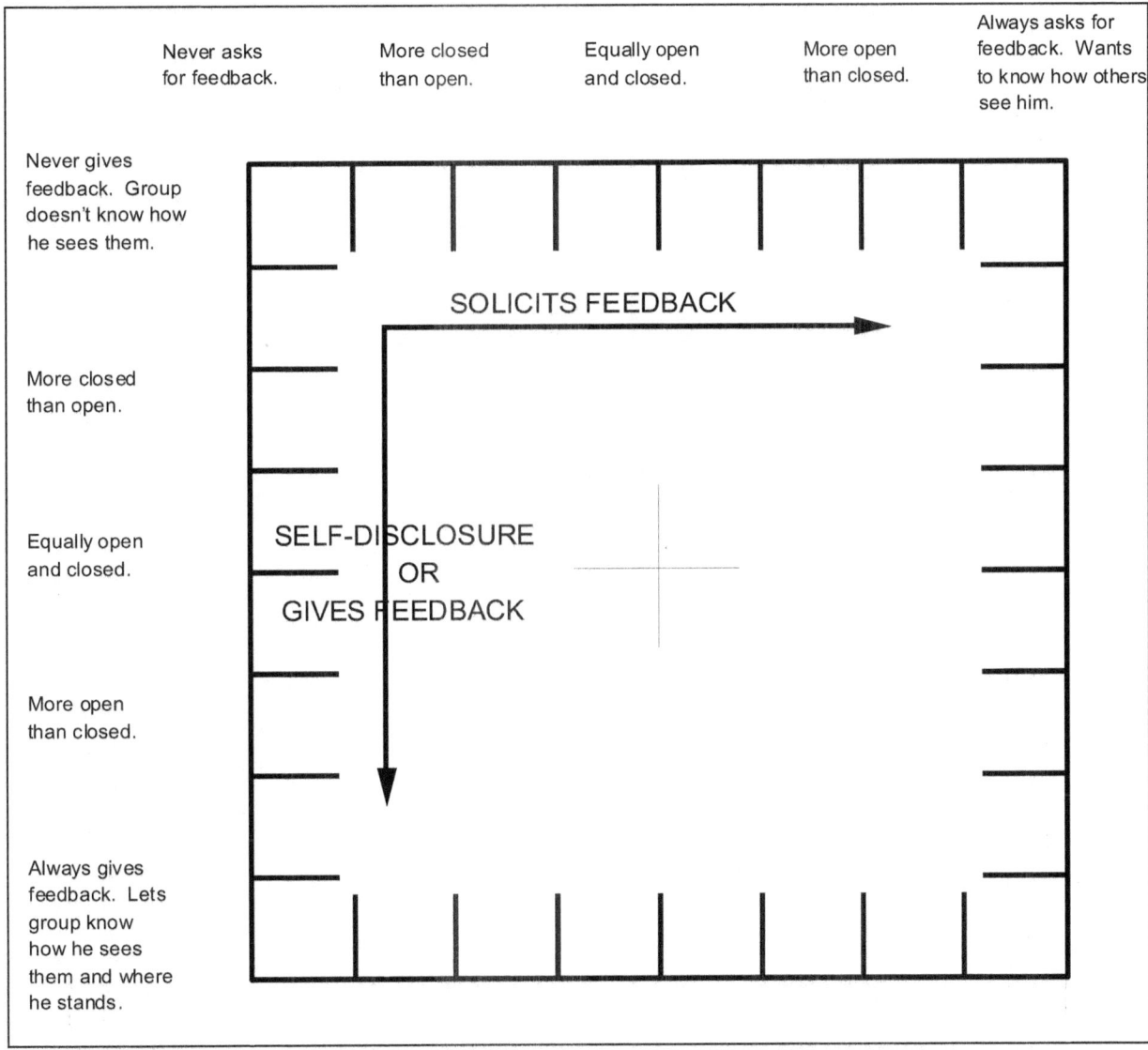

Adapted from <u>Group Processes: An Introduction to Group Dynamics</u> by Joseph Luft. Used with permission of Mayfield Publishing Co., Copyright 1984, 1970, and 1963 by Joseph Luft.

NOT USED

SMALL GROUP INSTRUCTOR TRAINING COURSE

(SGITC)

TRAINING SUPPORT PACKAGE FOR

LESSON 4:

EXPERIENTIAL LEARNING CYCLE

This page intentionally left blank.

Small Group Instruction

LESSON 4
EXPERIENTIAL LEARNING CYCLE

L4-A1

Small Group Instruction

TERMINAL OBJECTIVE

Explain the Experiential Learning Cycle (ELC).

L4-A2

Small Group Instruction

ENABLING OBJECTIVE 1

Define process and content as it relates to SGI.

L4-A3

Small Group Instruction

PROCESS vs CONTENT

Content = What was said.

Those topics discussed.

The task completed.

What was learned.

Process = How things were said.

How tasks were completed.

How emotions affected results.

How learning occurred.

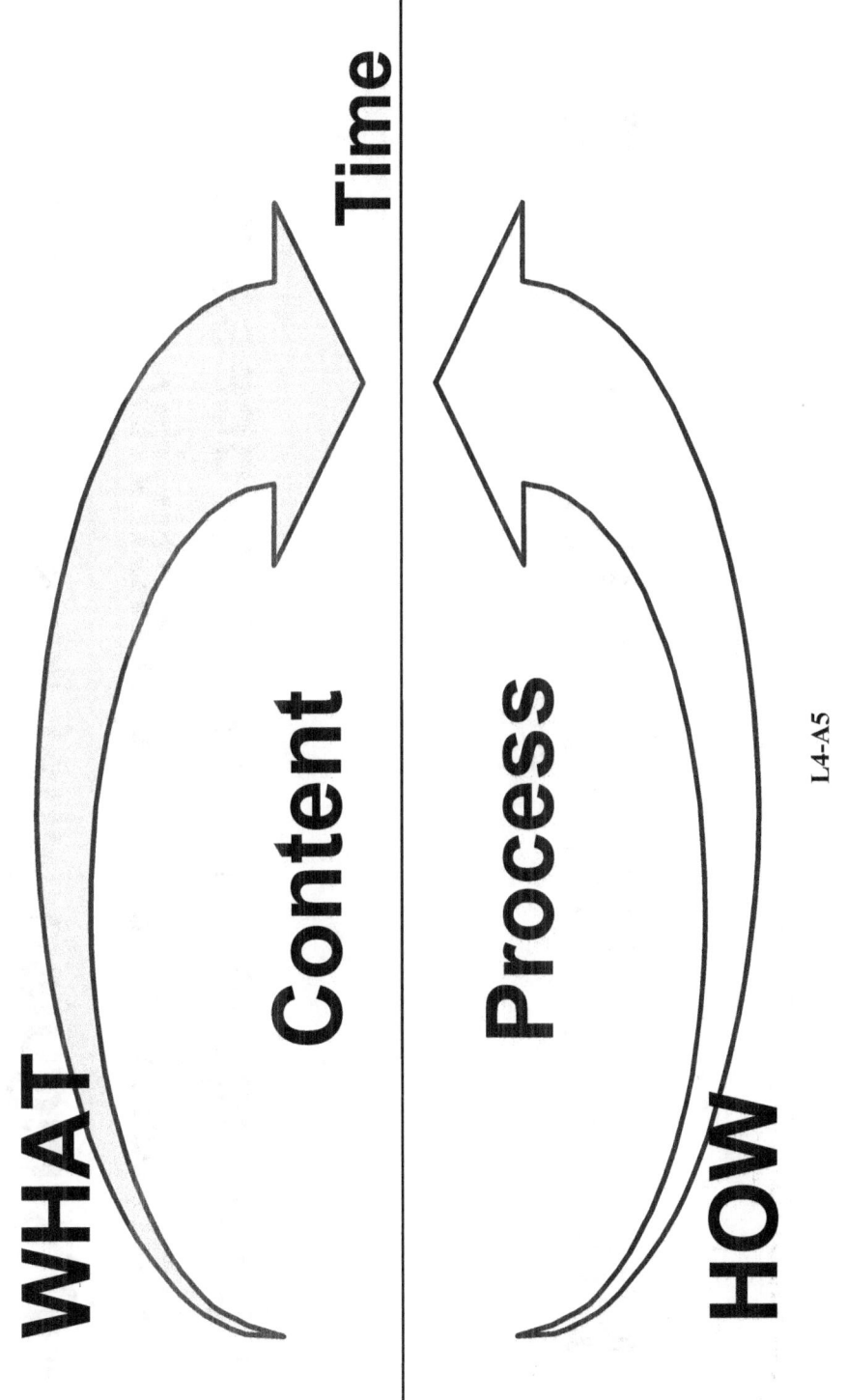

Small Group Instruction

ENABLING OBJECTIVE 2

- Distinguish between process and content in a group activity.
- Distinguish what was done from how it was done.

Small Group Instruction

PRACTICAL EXERCISE

L4-A7

Small Group Instruction

ENABLING OBJECTIVE 3

Describe the five phases of the Experiential Learning Cycle.

L4-A8

Small Group Instruction

EXPERIENTIAL LEARNING MODEL

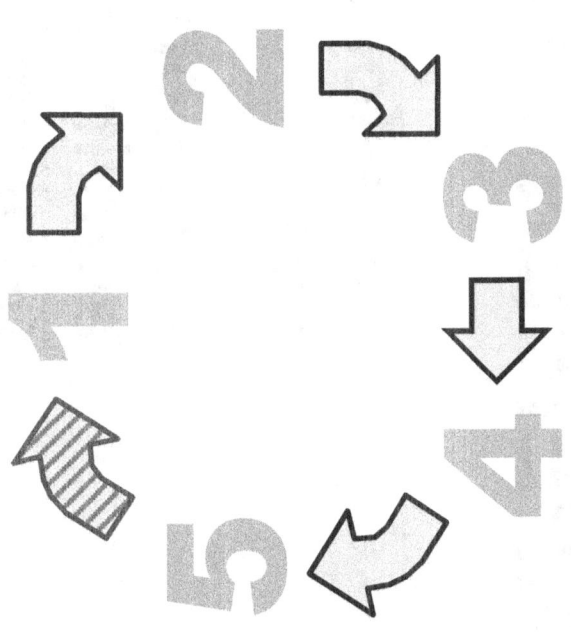

Small Group Instruction

EXPERIENCING

WHEN: Time focus for the event.

WHERE: Location of the activity or application of learning.

WHY, HOW, or PURPOSE: Rationale for lesson learned and what to do with it.

L4-A10

Small Group Instruction

EXPERIENCING
Here and now
In the group
Actions/tasks

Small Group Instruction

PUBLISHING

Here and now
In the group
Stating what occurred

E → P
↑   ↓
A ← G

L4-A12

Small Group Instruction

PROCESSING

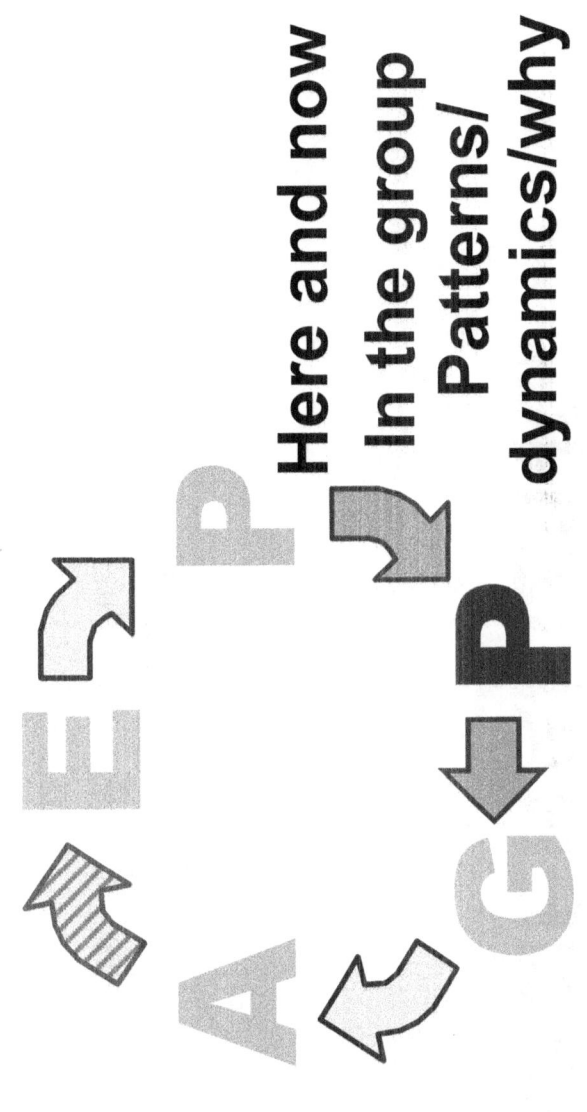

- Here and now
- In the group
- Patterns/dynamics/why

Small Group Instruction

GENERALIZING

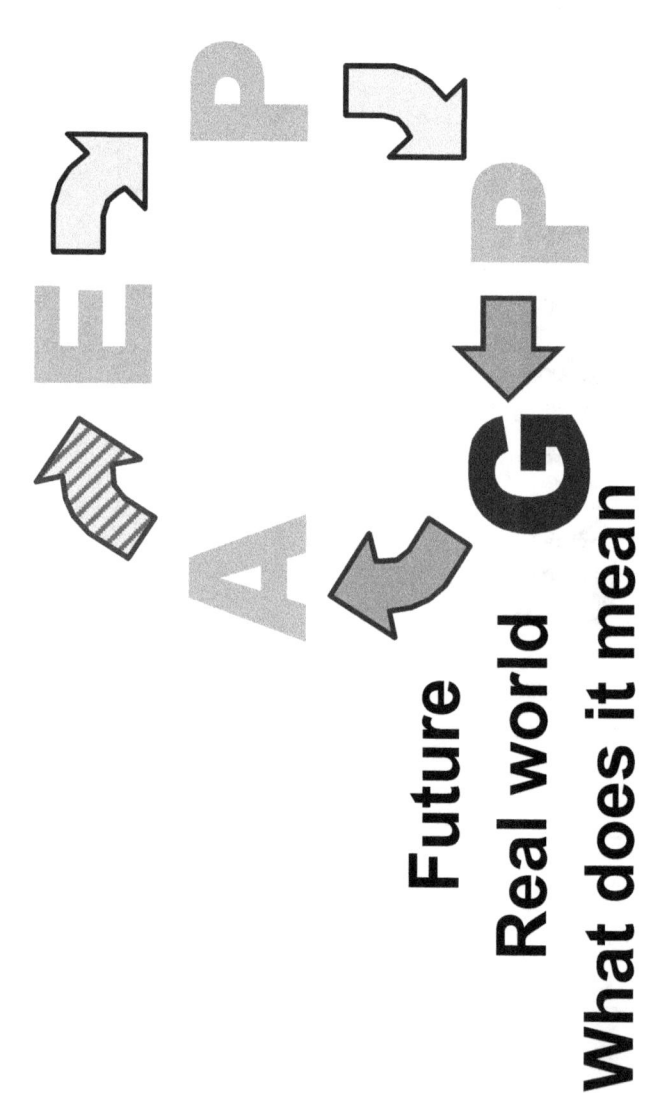

Future
Real world
What does it mean

L4-A14

Small Group Instruction

APPLYING

- Future
- Possible uses
- Improve future experience

E → A P P G ← A

L4-A15

Small Group Instruction

TRANSITIONS

Content to process is usually most difficult.

Small Group Instruction

TRANSITIONS

Individual reactions to group dynamics and here and now to real world.

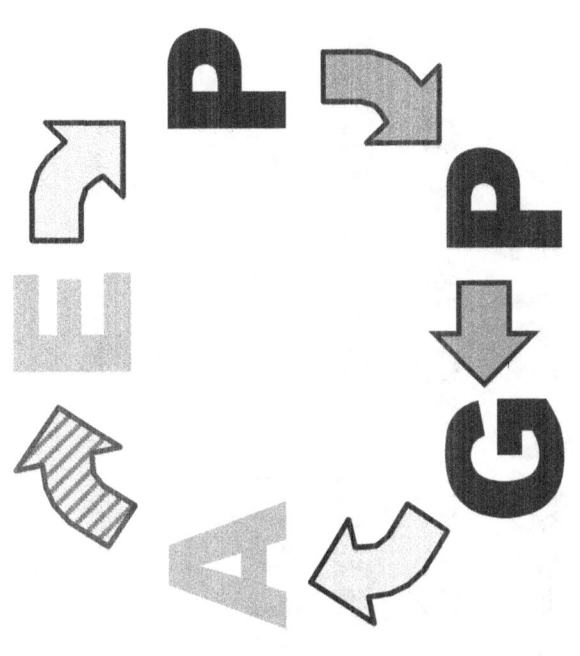

L4-A17

Small Group Instruction

TRANSITIONS

Transfer experience from real world to future or new activity to start cycle again.

L4-A18

Small Group Instruction

ENABLING OBJECTIVE 4

Compare the ELC to the AAR.

Include each AAR phase and the ELC stage that corresponds.

Small Group Instruction

ELC/AAR COMPARISON

- **Experiencing** — Training event
  - *Publishing* — *What occurred*
- **Processing** — What was right
  - *Generalizing* — *Next time*
- **Applying** — More training

L4-A20

Small Group Instruction

LESSON REVIEW

Content = WHAT is being learned.

Process = HOW groups learn.

L4-A21

Small Group Instruction

LESSON REVIEW

Experiential Learning Cycle:

- **Experiencing = What is the task?**
- **Publishing = What occurred?**
- **Processing = Why?**
- **Generalizing = What does it mean?**
- **Applying = Now what?**

L4-A22

Small Group Instruction

LESSON REVIEW

AAR = Looks at training content.

ELC = Looks also at social and psychological interactions.

L4-A23

NOT USED

LESSON 4

Student Handouts and Exercises

Handout	**Page**
The Experiential Learning Model (Blank)	SG4-27
The Experiential Learning Model (Complete)	SG4-29
ELC Questions	SG4-31

NOT USED

SGITC Lesson 4 - Experiential Learning Cycle

STUDENT HANDOUT

The Experiential Learning Model (Blank)

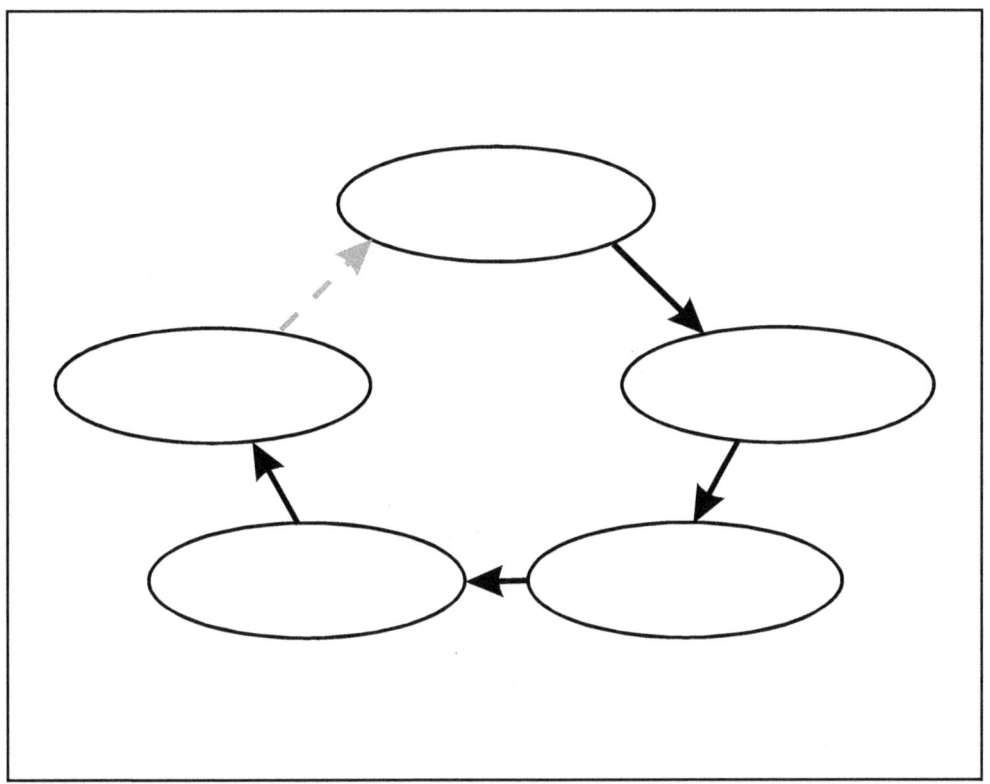

NOT USED

SGITC   Lesson 4 - Experiential Learning Cycle

STUDENT HANDOUT

The Experiential Learning Model

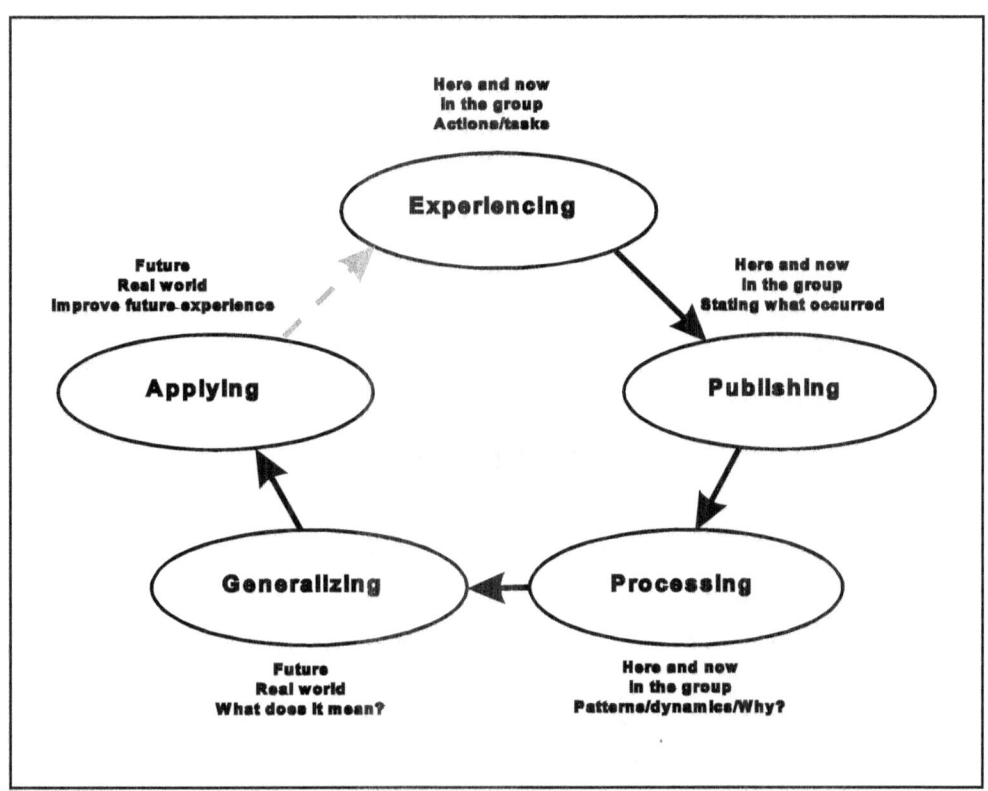

NOT USED

SGITC Lesson 4 - Experiential Learning Cycle

ELC Questions Handout

**PROCESSING QUESTIONS FOR EACH STAGE OF THE
EXPERIENTIAL LEARNING CYCLE**

Usually in stage one, the experiencing phase, group members participate in an activity to generate data. Processing the data does not actually begin until the second (publishing) stage. However, since group members sometimes resist beginning and/or completing an activity, the group leader may find the following questions helpful in stage one. They are usually "no fail" questions because (1) they tend to break down the group members' resistance by encouraging involvement in the activity; (2) if they do not break down the resistance, then processing this resistance becomes the learning; and (3) we can use them at any stage of the experiential cycle. They are key questions, which the group leader can use along with summarizing and reflecting to help the group move either more deeply into the stage at hand or on to another stage.

- What is going on?
- How do you feel about that?
- What do you need to know to _____?
- Would you be willing to try?
- Could you be more specific?
- Could you offer a suggestion?
- What would you prefer?
- What are your suspicions?
- What is your objection?
- If you could guess at the answer, what would it be?
- Can you say that in another way?
- What is the worst/best that could happen?
- What else?
- Would you say more about that?

Stage Two-Publishing

In stage two, publishing, group members have completed the experience. Questions focus on generating data.

- Who would volunteer to share? Who else?

- What happened?

- How did you feel about that?

- Who else had the same experience?

- Who reacted differently?

- Were there any surprises?

- How many felt the same?

- How many felt differently?

- What did you observe?

- What were you aware of?

Stage Three-Processing

In stage three, processing, group members now have data. Questions focus on making sense of that data for the individual and the group.

- How did you account for that?
- What does that mean to you?
- How was that significant?
- How was that good/bad?
- What struck you about that?
- How do those fit together?
- How might it have been different?
- Do you see something operating there?
- What does that suggest to you about yourself/your group?
- What do you understand better about yourself/your group?

Stage Four-Generalizing

In stage four, generalizing, group members work towards forming principles, which they derived from the specific knowledge they have gained about themselves and their group. Questions focus on promoting generalizations.

- What might we draw/pull from that?

- Is that plugging into anything?

- What did you learn/relearn?

- What does that suggest to you about _____ in general?

- Does that remind you of anything?

- What principle/law do you see operating?

- What does that help explain?

- How does this relate to other experiences?

- What do you associate with that?

- So what?

Stage Five-Applying

In stage five, applying, group members discuss using what they learned in their real-world situation. Questions focus on applying the general knowledge they have gained to their personal and professional lives.

- How could you apply/transfer that?
- What would you like to do with that?
- How could you repeat this again?
- What could you do to hold on to that?
- What are the options?
- What might you do to help/hinder yourself?
- How could you make it better?
- What would be the consequences of doing/not doing that?
- What modifications can you make work for you?
- What could you imagine/fantasize about that?

NOT USED

SMALL GROUP INSTRUCTOR TRAINING COURSE

(SGITC)

TRAINING SUPPORT PACKAGE FOR

LESSON 5:

INTERVENTION

This page intentionally left blank.

Small Group Instruction

LESSON 5
INTERVENTION

L5-A1

Small Group Instruction

TERMINAL OBJECTIVE

Identify how interventions can expedite the learning process in small groups.

Small Group Instruction

ENABLING OBJECTIVE 1

Identify the types of intervention and the objective, timing, and form of each.

Small Group Instruction

INTERVENTION TYPES

- Conceptual-input
- Coaching
- Process-observation

L5-A4

Small Group Instruction

INTERVENTION STRATEGIES

- Guidance
- Positive reinforcement
- Protection of an individual

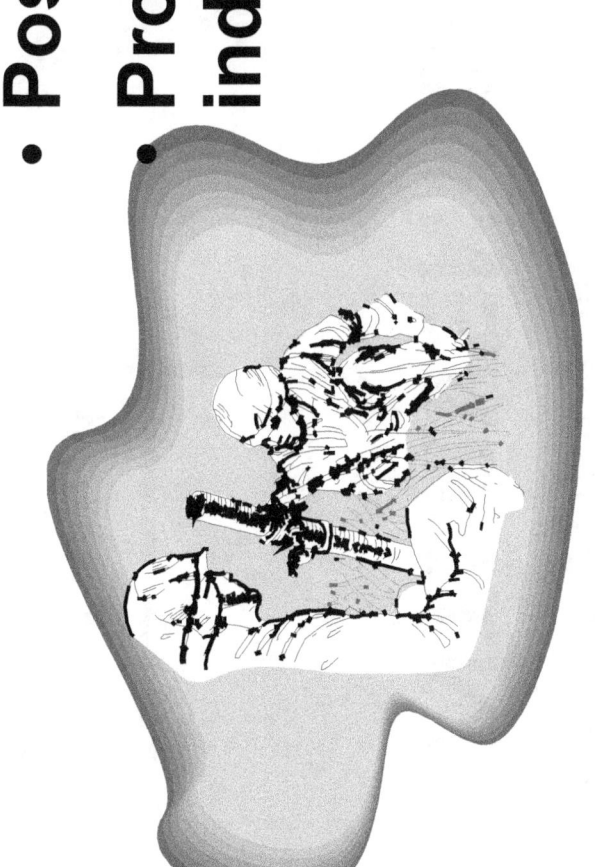

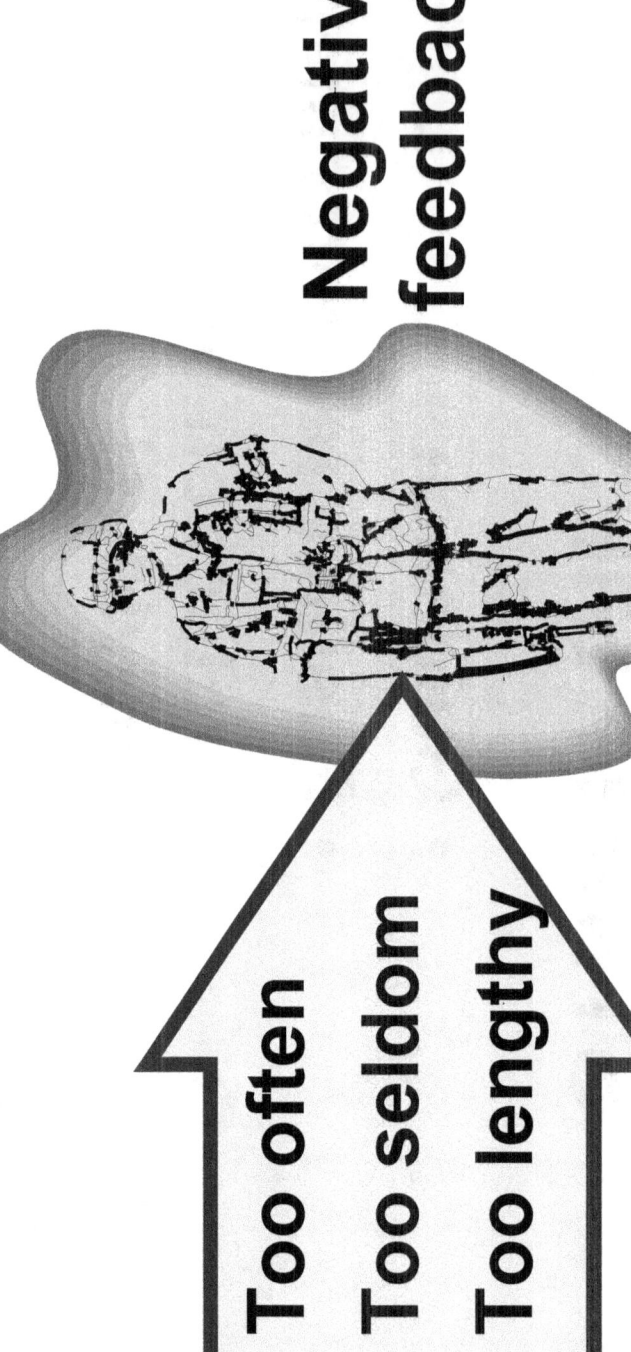

Small Group Instruction

SGL ACTIONS

- Step 1 - Diagnose.
- Step 2 - Identify role.
- Step 3 - Decide to act or to do nothing.

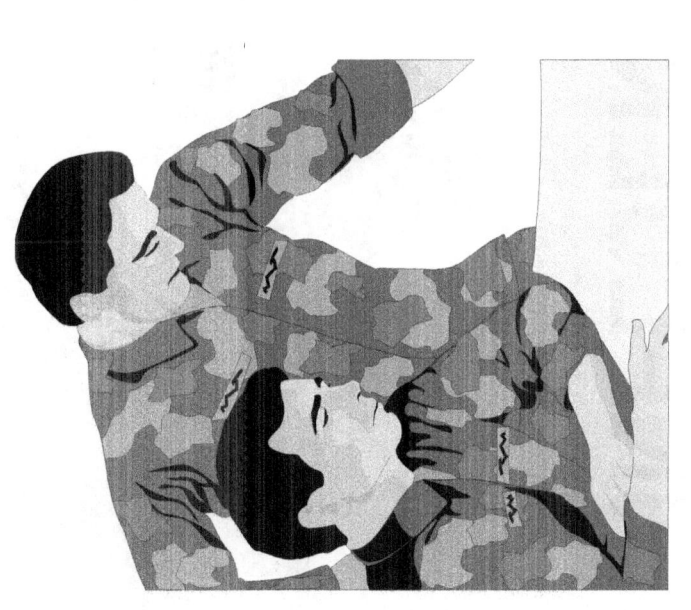

Small Group Instruction

INFLUENCING CONTENT

- Be specific.
- Be brief.
- Move group off center.

Small Group Instruction

INFLUENCING PROCESS

- Mention observations.
- Focus on process.
- Stimulate group interaction.

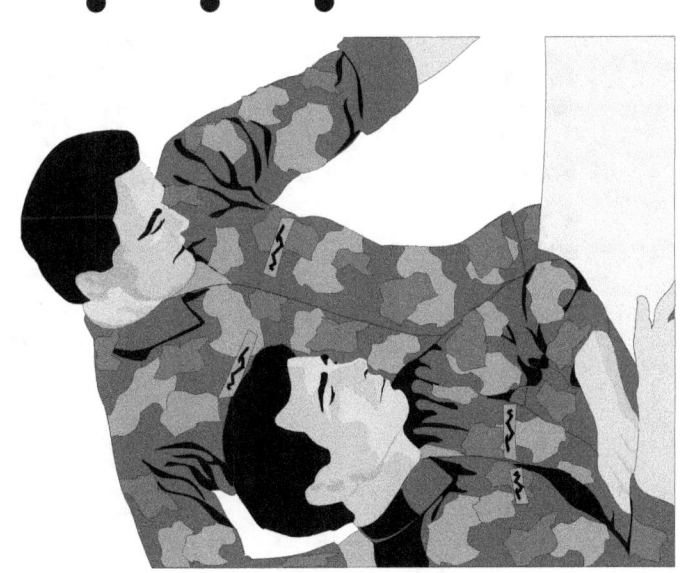

L5-A9

Small Group Instruction

ENABLING OBJECTIVE 2

Identify the nonproductive behavior that you will most likely encounter and how to deal with each behavior.

Small Group Instruction

REVIEW

- Three types of intervention
- Intervention strategies
- Intervention pitfalls
- Types of nonproductive behavior

NOT USED

LESSON 5

Student Handouts

Handout	Page
Situation 1 - Instructor Qualifications	SG5-15
Situation 2 - War Stories and "Bull Sessions"	SG5-17
Situation 3 - Nonparticipation	SG5-19
Situation 4 - Late Student	SG5-21
Situation 5 - The "Angry Huff"	SG5-23
Situation 6 - Les Miserables	SG5-25
Situation 7 - The Filibuster	SG5-27
Situation 8 - The "Corrector"	SG5-29
Situation 9 - "I Agree"	SG5-31
Situation 10 - Critique	SG5-33
Situation 11 - Attack	SG5-35
Situation 12 - Delay	SG5-37
Situation 13 - Self-worth	SG5-39
Situation 14 - Group Norms	SG5-41
Situation 15 - Adam Ant	SG5-43

NOT USED

SGITC   Lesson 5 – Intervention

STUDENT HANDOUT

Situation 1 - Instructor Qualifications

Situation It is the morning of the first day. The session has been in progress for approximately one and one-half hours. You are just processing one of the initial exercises in the session when a group member confronts you with the question of your qualifications to conduct this type of training. There is hostility in her tone of voice, and her demeanor indicates to you that there is more to her question than just a request for information. Even though it is only the first day, she has already expressed her displeasure with just about everything that has happened.

Question 1 Is the student displaying a nonproductive behavior? If so, which one?

Question 2 What would you do?

1. Nothing; ignore her behavior.

2. Reflect back on his question, and attempt to clarify her motivation.

3. Determine how other group members are reacting to her behavior.

4. Provide a structure wherein all group members can share their experience and qualifications in this area.

5. Describe this behavior as a form of resistance to authority of group leaders.

6. Confront her with your reactions to being challenged in a basically dishonest way.

7. Explain your qualifications and your authority to conduct the seminars.

8. Say "Questions about qualifications are trivial because they are not relevant to the here and now."

9. Comment that you are just one member of the group and that all members share the responsibility for making this a success.

10. Other. Explain.

NOT USED

SGITC   Lesson 5 – Intervention

STUDENT HANDOUT

Situation 2 - War Stories and "Bull Sessions"

Situation	The group has returned from lunch the first day. The group members are talking about their past experiences. War stories and "Bull sessions" are going on. They seem to want to get back to work, but one student is preventing this by bringing up extraneous material and comments/questions not related to the topic.
Question 1	Is the student displaying a nonproductive behavior? If so, which one?
Question 2	What would you do?

1. Nothing.

2. Participate in the discussion, and try to turn it back to the topic at hand.

3. Point out the irrelevancy of the current discussion.

4. Share your feelings on how the individual is acting.

5. Provide an activity that focuses the group back into the here and now.

6. State that you see the current group behavior as a means to avoid dealing with the here and now.

7. Share your reactions to an obvious violation of the "here and now" ground rules for the group and how this is affecting you as the facilitator.

8. Ask the group if they think the discussion is relevant to the goals of the course.

9. Ask group members how they feel about participating in this type of discussion.

10. Other. Explain.

NOT USED

SGITC   Lesson 5 – Intervention

STUDENT HANDOUT

Situation 3 - Nonparticipation

Situation 3 It is the morning of the second day. The session has been in progress for approximately 1 hour. You have just initiated a feedback exercise and notice that an individual has not participated in the exercise. The student appears to be elsewhere mentally and has repeatedly asked you to repeat questions/directions.

Question 1 Is the student displaying a nonproductive behavior? If so, which one?

Question 2 What would you do?

1. Nothing.

2. State that silence is often productive in groups of this nature.

3. Solicit other group members' responses to this student's prolonged silence.

4. Share your feelings as facilitator about the nonparticipation.

5. Structure an additional activity designed to encourage more participation.

6. Describe the avoidance behaviors you are currently observing in the group.

7. Question the student and solicit reasons for nonparticipation.

8. Re-emphasize that everyone is expected to participate in this activity.

9. Point out to the group in general that you are aware of how hard it is to share one's feelings.

10. Other. Explain.

NOT USED

SGITC   Lesson 5 – Intervention

STUDENT HANDOUT

Situation 4 - Late Student

Situation   It is the afternoon of the second day. The group has just reconvened after a one and a half hour lunch break. One group member returns to the group approximately 30 minutes late. He has been late before, and is habitually late returning from coffee breaks, but no one has said or done anything about it. He also has been talking while others are talking and asks questions in a hostile manner.

Question 1   Is the student displaying a nonproductive behavior? If so, which one?

Question 2   What do you do?

1. Nothing. Ignore the behavior.

2. Solicit other group members' feelings about the behavior of the late person.

3. Share your feelings about members coming late.

4. Suggest that the group member try to get to sessions on time.

5. Comment that lateness is often an attention getting ploy.

6. Share your anger, and describe the disruptive effect that lateness has on the activity now in progress.

7. Describe how this member's returning late seems to have affected the behavior of other group members.

8. Ask him how he feels about it being brought up that he is late.

9. Comment that his behavior is a violation of the ground rules and that perhaps he should not be in the course if he has so many other things that keep him from being on time.

10. Other. Explain.

NOT USED

SGITC   Lesson 5 – Intervention

STUDENT HANDOUT

Situation 5 – The "Angry Huff"

Situation 5	It is late in the afternoon of the second day. Two group members were just involved in a very intense confrontation, and one of them left the room quite suddenly in an angry huff. A third group member seems upset by the actions of the other two and attempts to smooth things over, apologizing for the actions of the student that left.
Question 1	Is the third student displaying a nonproductive behavior? If so, which one?
Question 2	What do you do?

1. Nothing.

2. Say that it's okay for them to be angry with you.

3. Invite the group member who has just apologized to you to work through the issue.

4. Say that the departing member is going to have to speak for himself.

5. Describe what is going on in the group and how this relates to the previous confrontation.

6. Question the group member about his reasons for dealing with you in this fashion.

7. Turn the group over to your teammate. Meet with the departing member and discuss the inappropriateness of his behavior.

8. Go and talk to the departing member, and use active listening to help him clarify the underlying issue.

9. Tell the remaining member involved in the original conflict that _he_ has responsibility for resolving his conflict, and the group will wait for his return.

10. Other. Explain.

NOT USED

SGITC Lesson 5 – Intervention

STUDENT HANDOUT

Situation 6 - Les Miserables

Situation It is the morning of the third day. One of the members has said very little throughout the meeting and seems to be miserable. Several times this member has had a look of disgust or pain. Some of the others are beginning to question this member about the silence. The member remains quiet, and the group seems uncertain about how to proceed.

Question 1 Is this student displaying a nonproductive behavior? If so, which one?

Question 2 What do you do?

1. Even though they look to you for help, leave it to the group to deal with the situation.

2. Comment how the silent group member's behavior is affecting the operation of the group.

3. Ask others to share their feelings about the group member's silence.

4. Provide a structure that allows the group member to participate more actively in the group and share feelings.

5. Discuss the dynamics of low participation by commenting that silence is most likely a consequence of anxiety at being involved in this type of training.

6. Question the group member about why the member has not participated more freely in the group.

7. Facilitate the communications between the silent member and one of those confronting the silent member so that the silent member can be drawn into the group.

8. Intervene in the confrontation, and comment that silence is okay.

9. Intervene in the confrontation, and point out that one of the ground rules is that all are responsible for their own level of participation and that members will get out of the group what they put into it.

10. Other. Explain.

NOT USED

SGITC   Lesson 5 – Intervention

STUDENT HANDOUT

Situation 7 - The Filibuster

Situation	It is now the afternoon of the third day and the conversation has been monopolized by one of the group members. His monologue and interruptions have interfered with the development of the group and blocked any kind of meaningful interchange. During this session, he has had the floor for approximately 1 hour.
Question 1	Is the student displaying a nonproductive behavior? If so, which one?
Question 2	What do you do?

1. Even though the group members seem to be frustrated and look to you for help, leave it to the group to deal with the situation.

2. Ask other group members to describe what has been going on in this session.

3. Ask group members how they feel about one person doing most of the talking.

4. Share your feelings about the monopolization of the group.

5. Direct your remarks to other group members in an attempt to increase their participation.

6. Describe what has been going on as primarily a two-party interaction, where one has monopolized the conversation while other group members have allowed this type of behavior.

7. Tell the member how much time he has been controlling the discussion and to give others a chance.

8. Confront him (using effective communications) with your feelings about his participation and then actively listen to his response.

9. Wait until you observe another member's agitation, and facilitate a confrontation between the two.

10. Other. Explain.

NOT USED

SGITC   Lesson 5 – Intervention

STUDENT HANDOUT

Situation 8 - The "Corrector"

Situation	It is the morning of the fourth day. Approximately one-half hour into the day, one of the group members announces that he is going to quit the group. He says that he cannot see how the training is of benefit to him since he has not learned anything he didn't already know. He has added to or corrected everything that has been said to this point, many times giving incorrect information. Other group members are upset by his announcement and try to talk him out of leaving.
Question 1	Is the student displaying a nonproductive behavior? If so, which one?
Question 2	What do you do?

1. Nothing.

2. Say that you have enjoyed his participation in the group and would be sorry if he left.

3. Ask group members what they should do about the situation.

4. Share your feelings about this group member leaving.

5. Remind the group member that he is there under orders and that before you allow him to leave he will have to state his reasons more fully.

6. Invite him to sit down and use active listening techniques to help him verbalize his feelings.

7. Provide a structure that will allow group members to voice their feelings about this member's actions.

8. Describe how this behavior seems to have everyone concerned and confused as to what to do.

9. Ask him to participate with an open mind until the end of the day. Then if he still wants to leave, he may.

10. Other: Explain.

NOT USED

STUDENT HANDOUT

Situation 9 – "I Agree"

Situation	It is the morning of the fifth day. The group is doing well, and members are taking responsibility for maintaining group effectiveness and the ground rules. One of the students has been participating regularly, but you are starting to notice that she never disagrees with anyone and seems to be avoiding any type of conflict. She is causing no problems, but you are concerned about her growth since she has given nothing but glowing feedback.
Question 1	Is the student displaying a nonproductive behavior? If so, which one?
Question 2	What do you do?

1. Do nothing.

2. Continue the class, letting her take responsibility for her feelings.

3. Use active listening techniques to help her express her feelings.

4. Provide a structure, which allows other members to express how they feel about her lack of conflict.

5. Tell her that you want her to share all her feelings, not just the positive ones.

6. Ask members of the group to diagnose what state of group development this situation signifies.

7. So that she won't feel embarrassed, make a general comment to the group that expressing dissent and negative reactions is part of the group growing process.

8. Other. Explain.

NOT USED

SGITC   Lesson 5 – Intervention

STUDENT HANDOUT

Situation 10 - Critique

Situation The workshop is drawing to a close and the participants are conducting a critique of it. You have requested feedback about its value to each individual member. One participant, who has been a low participator throughout, comments that it was a waste of time. He had interrupted you repeatedly throughout the class, trying to get you to reveal answers instead of figuring them out for himself.

Question 1 Which nonproductive behavior is this student displaying?

Question 2 What do you do?

1. Remain silent.

2. Interrupt the student, and state that the purpose of the critique is for constructive feedback.

3. State that it is okay for people to have negative feelings about the workshop.

4. Ask other participants how they feel about the student's constant interruptions.

5. Ask the member why he feels he should have all his questions answered instead of first trying to figure them out himself.

6. Comment that this situation indicates that the group did not fully understand the ground rules and its self-control responsibilities.

7. Other. Explain.

NOT USED

SGITC Lesson 5 – Intervention

STUDENT HANDOUT

Situation 11 - Attack

Situation It is the morning of the third day. The group has worked through Stage 1 issues and is exhibiting Stage 2 behavior. At this point, one of the group members (who has continually blocked group process, withdrawn, baited another group member, and then withdrawn again) leads an attack against you, the facilitator, by accusing you of not doing your job. He thinks you should do more instructing and involve the nonparticipants.

Question 1 Is the student displaying a nonproductive behavior? If so, which one?

Question 2 What do you do?

1. Ask the group what it thinks.

2. Let the group take responsibility for policing itself.

3. Defend your role and style as a facilitator.

4. Do not attack the attacker. Provide him some feedback about his behavior and its effect on the group.

5. Do nothing. Who cares what he thinks.

6. Other. Explain.

NOT USED

SGITC Lesson 5 – Intervention

STUDENT HANDOUT

Situation 12 - Delay

Situation It is the end of the third week. The group is progressing well and accomplishing all tasks. One member is delaying all activities, however, because he tends to read complex theories into simple thoughts. He takes every word literally and tries to make sense out of everything. The rest of the group is beginning to resent his actions.

Question 1 What nonproductive behavior is this person exhibiting?

Question 2 What do you do?

1. Take charge of the group and jump right in to the next task without allowing him to delay.

2. Allow the group members to take care of the situation themselves.

3. Take the member aside and counsel him on how his behavior is affecting the rest of the group.

4. Prior to the class, pick two of the best liked students of the group and ask them to take charge of preventing the individual from continuing.

5. Next time he becomes verbose, tell him to try to express himself in simpler terms (i.e., as a one sentence idea).

6. Other. Explain.

NOT USED

SGITC   Lesson 5 – Intervention

STUDENT HANDOUT

Situation 13 - Self-worth

Situation   It is the morning of the fourth day. The group is still in Stage 1 behavior, but members are starting to exert some initiative and independence. Exercises have been slow but completed with passable results. Students are starting to feel more at home in the group, interjecting as appropriate and actively participating in activities. One member is apparently having some self-worth problems and constantly begins each question or statement with an explanation/apology.

Question 1   What behavior is the student expressing?

Question 2   What do you do?

1. Nothing.

2. Confront the individual in the group saying that he has contributed to the class and doesn't have to apologize every time he speaks.

3. Counsel the individual in private.

4. Say that you feel bad that the individual feels as though he must apologize every time he speaks.

5. Let the group deal with the issue if it bothers them.

6. Research the individual's background to see why he feels inadequate. Then try to structure an activity that will allow him to shine as the expert.

7. Other. Explain.

NOT USED

SGITC Lesson 5 – Intervention

STUDENT HANDOUT

Situation 14 - Group Norms

Situation It is the afternoon of day four. You learned at lunch, from a member of your group, that another member has some very deep feelings about the agreed upon group norms. The first member relates that during lunch the second member expressed negative feelings about the self-governing rules that were established. Because of the generalities he is using, you suspect that the first member may really be having those feelings himself.

Question 1 Is the first member expressing a nonproductive behavior? If so, which one?

Question 2 What do you do?

1. Nothing.

2. Tell the group what you have heard, and let them deal with it.

3. Counsel him (in private) and see if he is really the disgruntled member. Determine if the second member does have problems, or if both of them have problems.

4. Ask him why he is attributing his feelings to other people.

5. While in the group ask him, "Is that a statement from the group, or is it personal?"

6. Wait until there is a group impact.

7. Other. Explain.

NOT USED

SGITC   Lesson 5 – Intervention

STUDENT HANDOUT

Situation 15 - Adam Ant

Situation	It is the third day. You have subdivided the group into two smaller groups and given them a problem-solving task that requires them to reach group consensus. Voting and majority decisions are forbidden by the rules of the exercise. One of the members, a productive participant up to this point, has suddenly become adamant about an issue that the rest of the group cannot agree to. The allotted time has expired, and the group containing the member is not anywhere near a consensus solution to the task.
Question 1	Has anyone exhibited a nonproductive behavior? If so, which one?
Question 2	What do you do?

1. Give them more time to reach consensus and solve the problem.

2. Change the rules, and allow voting and majority decisions.

3. Stop what they are doing, and proceed to the next lesson.

4. Conduct an after action review.

5. Process the activity, both what they did and how they did it.

6. Other. Explain.

NOT USED

SMALL GROUP INSTRUCTOR TRAINING COURSE

(SGITC)

TRAINING SUPPORT PACKAGE FOR

LESSON 6:

LEADERLESS DISCUSSIONS

This page intentionally left blank.

Small Group Instruction

LESSON 6
LEADERLESS DISCUSSIONS

Small Group Instruction

LEADERLESS DISCUSSIONS:

The Brainstorm

Small Group Instruction

TERMINAL OBJECTIVE

Conduct a brainstorming session.

Small Group Instruction

RULES FOR THE GENERATING PHASE OF BRAINSTORMING

- There will be no criticism.
- Far-fetched ideas are desirable.
- Many ideas help the process.
- Flip-flop technique is helpful.
- Piggyback technique is useful.

L6a-A4

Small Group Instruction

RULES FOR THE EVALUATING PHASE OF BRAINSTORMING

- Criticism of list is authorized.
- Answer question/meet objective.
- Attain group consensus.

Small Group Instruction

WHAT A TEAM LOOKS LIKE

Top Ten:

L6a-A6

Small Group Instruction

LESSON REVIEW

A Brainstorm:

- **Generates creative solutions to problems**
- **Has two distinct phases**
  - Generalizing
  - Evaluating

L6a-A7

NOT USED

LESSON 6a

Student Handouts

Handout	Page
Rules for the Generating Phase of Brainstorming	SG6a-11
Rules for Evaluating Phase of Brainstorming	SG6a-13
Performance Evaluation Checklist	SG6a-15

NOT USED

STUDENT HANDOUT

Rules for the Generating Phase of Brainstorming

- No criticism of ideas during generation phase.

- Far-fetched ideas are helpful.

- Many ideas are desirable.

- Flip-flop technique is helpful.

- Piggyback technique is helpful.

NOT USED

STUDENT HANDOUT

Rules for the Evaluating Phase of Brainstorming

- Criticism of topic is authorized.

- Answer question/meet objective.

- Attain group consensus.

NOT USED

SGITC   Lesson 6a – Leaderless Discussions: The Brainstorm

Performance Evaluation Checklist

Purpose	The PEC is used to assess the SDL and the small group during the SDL presentation of an SGI lesson and final exam.
Who uses this checklist	The following personnel use this checklist: • Observing groups at the SDL site to rate the group and the SDL. • SGL to rate the SDL and group performance.
How to use this checklist	a. Observing groups: Fill out one checklist for the group. Discuss your observations and come to consensus. Ensure your discussion does not disturb the SDL's presentation. Look for positive and correctable SDL attributes and group dynamics. Try to link SDL roles, group process and content issues, and group dynamics to the success of the group meeting the training objective. Report out after the SDL reviews the ELC. b. SGL: Fill out checklist based on your observations of group and SDL performance.
When to use this checklist	This checklist is used for student lesson presentations (lessons 6-10) and the final exam.

NOT USED

SGITC   Lesson 6a – Leaderless Discussions: The Brainstorm

PERFORMANCE EVALUATION CHECKLIST (Student Discussion Leader Rating)

Student Name: _____   Date: _____

Evaluator: _____   Rating: _____

Instructions: Evaluate student delivered lessons (6-10) and the final examination with this checklist. Students must receive the number of "GOs" listed next to the evaluated section, i.e., Introduction, 3 "GOs" out of 4. Students must pass each section (Introduction, Body, Conclusion, and ELC) to pass the course. Circle a GO or NO GO for each element in a section and the corresponding rating for that section. Write student rating (a GO or NO GO) on the Rating line above.

TOPIC		COMMENTS
Introduction (3 of 4). Did SDL:	**Go/ No Go**	
Focus group on task?	Go/ No Go	
Clearly state objective?	Go/ No Go	
Motivate/interest students?	Go/ No Go	
Tie SGI method to lesson objective?	Go/ No Go	
Lesson Body (3 of 3). Did SDL:	**Go/ No Go**	
Display SGL roles: - SME? - Facilitator? - Observer?	Go/ No Go	
Involve all students?	Go/ No Go	
Display understanding of SGI method chosen?	Go/ No Go	
ELC (4 of 4). Did SDL:	**Go/ No Go**	
Lead group in publishing stage?	Go/ No Go	
Lead group in processing stage?	Go/ No Go	
Lead group in generalizing stage?	Go/ No Go	
Lead group in applying stage?	Go/ No Go	
Conclusion (2 of 3). Did SDL:	**Go/ No Go**	
Review lesson objective and SGI method relationship?	Go/ No Go	
Summarize lesson results?	Go/ No Go	
Achieve training objective?	Go/ No Go	

Lesson 6a - Leaderless Discussions: The Brainstorm   SGITC

PERFORMANCE EVALUATION CHECKLIST (Group)

Lesson: _____   Date: _____

SDL: _____   Evaluator: _____

Instructions: This checklist is used to evaluate group performance during student-led lessons (6-10). Observers within and outside the classroom use the checklist. The SGL uses this form to comment on group productivity and development.

Group criteria do not effect students' evaluation for graduation.

TOPIC		COMMENTS
During the lesson did:		
Group members interact with each other?	Yes / No	
All group members participate?	Yes / No	
Group work as a team?	Yes / No	
Group members provide feedback?	Yes / No	
Each student publish during the ELC?	Yes / No	
Group adheres to the group rules?	Yes / No	
SDL interventions help the group?	Yes / No	
The group focuses on the task?	Yes / No	
Group achieves the training objective?	Yes / No	

Additional Comments:

Small Group Instruction

LESSON 6
LEADERLESS DISCUSSIONS

Small Group Instruction

LEADERLESS DISCUSSIONS:
The Buzz Session

Small Group Instruction

TERMINAL OBJECTIVE

Conduct a buzz session so that students gain:

- Introduction to issues or problems.
- Increased involvement in course content.
- Increased participation in class discussions.

Small Group Instruction

TOPIC #1: "What are the qualities of a good small group leader?"

- Discuss in your small group.
- Record your main points.
- Report back in 15 minutes.

Small Group Instruction

TOPIC #2: "What do good small group leaders expect from their students?"

- Discuss in your small group.
- Record your main points.
- Report back in 15 minutes.

L6b-A5

Small Group Instruction

TOPIC #3: "What do good small group leaders never do?"

- Discuss in your small group.
- Record your main points.
- Report back in 15 minutes.

Small Group Instruction

LESSON REVIEW

A Buzz Session:

- Introduces topics or problems.
- Encourages free discussion of issues.
- Increases participation in class discussions.

L6b-A7

NOT USED

LESSON 6b

Student Handout

Handout	**Page**
Performance Evaluation Checklist	SG6b-11

NOT USED

Performance Evaluation Checklist

Purpose	The PEC is used to assess the SDL and the small group during the SDL presentation of an SGI lesson and final exam.
Who uses this checklist	The following personnel use this checklist: • Observing groups at the SDL site to rate the group and the SDL. • SGL to rate the SDL and group performance.
How to use this checklist	a. Observing groups: Fill out one checklist for the group. Discuss your observations and come to consensus. Ensure your discussion does not disturb the SDL's presentation. Look for positive and correctable SDL attributes and group dynamics. Try to link SDL roles, group process and content issues, and group dynamics to the success of the group meeting the training objective. Report out after the SDL reviews the ELC. b. SGL: Fill out checklist based on your observations of group and SDL performance.
When to use this checklist	This checklist is used for student lesson presentations (lessons 6-10) and the final exam.

NOT USED

SGITC   Lesson 6b – Leaderless Discussions: The Buzz Session

PERFORMANCE EVALUATION CHECKLIST (Student Discussion Leader Rating)

Student Name: _____   Date:_____

Evaluator:_____   Rating: _____

Instructions: Evaluate student delivered lessons (6-10) and the final examination with this checklist. Students must receive the number of "GOs" listed next to the evaluated section, i.e., Introduction, 3 "GOs" out of 4. Students must pass each section (Introduction, Body, Conclusion, and ELC) to pass the course. Circle a GO or NO GO for each element in a section and the corresponding rating for that section. Write student rating (a GO or NO GO) on the Rating line above.

TOPIC		COMMENTS
Introduction (3 of 4). Did SDL:	**Go/ No Go**	
Focus group on task?	Go/ No Go	
Clearly state objective?	Go/ No Go	
Motivate/interest students?	Go/ No Go	
Tie SGI method to lesson objective?	Go/ No Go	
Lesson Body (3 of 3). Did SDL:	**Go/ No Go**	
Display SGL roles: - SME? - Facilitator? - Observer?	Go/ No Go	
Involve all students?	Go/ No Go	
Display understanding of SGI method chosen?	Go/ No Go	
ELC (4 of 4). Did SDL:	**Go/ No Go**	
Lead group in publishing stage?	Go/ No Go	
Lead group in processing stage?	Go/ No Go	
Lead group in generalizing stage?	Go/ No Go	
Lead group in applying stage?	Go/ No Go	
Conclusion (2 of 3). Did SDL:	**Go/ No Go**	
Review lesson objective and SGI method relationship?	Go/ No Go	
Summarize lesson results?	Go/ No Go	
Achieve training objective?	Go/ No Go	

Lesson 6b - Leaderless Discussions: The Buzz Session   SGITC

PERFORMANCE EVALUATION CHECKLIST (Group)

Lesson: _____   Date: _____

SDL: _____   Evaluator: _____

Instructions: This checklist is used to evaluate group performance during student-led lessons (6-10). Observers within and outside the classroom use the checklist. The SGL uses this form to comment on group productivity and development.

Group criteria do not effect students' evaluation for graduation.

TOPIC		COMMENTS
During the lesson did:		
Group members interact with each other?	Yes / No	
All group members participate?	Yes / No	
Group work as a team?	Yes / No	
Group members provide feedback?	Yes / No	
Each student publish during the ELC?	Yes / No	
Group adheres to the group rules?	Yes / No	
SDL interventions help the group?	Yes / No	
The group focuses on the task?	Yes / No	
Group achieves the training objective?	Yes / No	

Additional Comments:

Small Group Instruction

LESSON 6
LEADERLESS DISCUSSIONS

Small Group Instruction

LEADERLESS DISCUSSIONS:
Topic Discussion

L6c-A2

Small Group Instruction

TERMINAL OBJECTIVE

Conduct a topic discussion to provide students:

- Issue or problem introduction.
- Insight into possible solutions.
- Consequences of applying new methods or techniques.
- Learning reinforcement.

L6c-A3

Small Group Instruction

TOPIC: "What is the role of SGI in the Army and does it work?"

- Discuss in your small group.
- Report back in 15 minutes.

Small Group Instruction

LESSON REVIEW

A Topic Discussion:

- Introduces issues or problems.
- Reinforces learning.
- Increases group involvement.

NOT USED

SGITC

Lesson 6c – Leaderless Discussions: Topic Discussion

LESSON 6c

Student Handout

Handout | **Page**

Performance Evaluation Checklist SG6c-7

NOT USED

SGITC Lesson 6c – Leaderless Discussions: Topic Discussion

Performance Evaluation Checklist

Purpose	The PEC is used to assess the SDL and the small group during the SDL presentation of an SGI lesson and final exam.
Who uses this checklist	The following personnel use this checklist: • Observing groups at the SDL site to rate the group and the SDL. • SGL to rate the SDL and group performance.
How to use this checklist	a. Observing groups: Fill out one checklist for the group. Discuss your observations and come to consensus. Ensure your discussion does not disturb the SDL's presentation. Look for positive and correctable SDL attributes and group dynamics. Try to link SDL roles, group process and content issues, and group dynamics to the success of the group meeting the training objective. Report out after the SDL reviews the ELC. b. SGL: Fill out checklist based on your observations of group and SDL performance.
When to use this checklist	This checklist is used for student lesson presentations (lessons 6-10) and the final exam.

NOT USED

SGITC   Lesson 6c – Leaderless Discussions: Topic Discussion

PERFORMANCE EVALUATION CHECKLIST (Student Discussion Leader Rating)

Student Name: _____   Date: _____

Evaluator: _____   Rating: _____

Instructions: Evaluate student delivered lessons (6-10) and the final examination with this checklist. Students must receive the number of "GOs" listed next to the evaluated section, i.e., Introduction, 3 "GOs" out of 4. Students must pass each section (Introduction, Body, Conclusion, and ELC) to pass the course. Circle a GO or NO GO for each element in a section and the corresponding rating for that section. Write student rating (a GO or NO GO) on the Rating line above.

TOPIC		COMMENTS
Introduction (3 of 4). Did SDL:	**Go/ No Go**	
Focus group on task?	Go/ No Go	
Clearly state objective?	Go/ No Go	
Motivate/interest students?	Go/ No Go	
Tie SGI method to lesson objective?	Go/ No Go	
Lesson Body (3 of 3). Did SDL:	**Go/ No Go**	
Display SGL roles: - SME? - Facilitator? - Observer?	Go/ No Go	
Involve all students?	Go/ No Go	
Display understanding of SGI method chosen?	Go/ No Go	
ELC (4 of 4). Did SDL:	**Go/ No Go**	
Lead group in publishing stage?	Go/ No Go	
Lead group in processing stage?	Go/ No Go	
Lead group in generalizing stage?	Go/ No Go	
Lead group in applying stage?	Go/ No Go	
Conclusion (2 of 3). Did SDL:	**Go/ No Go**	
Review lesson objective and SGI method relationship?	Go/ No Go	
Summarize lesson results?	Go/ No Go	
Achieve training objective?	Go/ No Go	

Lesson 6c – Leaderless Discussions: Topic Discussion   SGITC

PERFORMANCE EVALUATION CHECKLIST (Group)

Lesson: _____   Date: _____

SDL: _____   Evaluator: _____

Instructions: This checklist is used to evaluate group performance during student-led lessons (6-10). Observers within and outside the classroom use the checklist. The SGL uses this form to comment on group productivity and development.

Group criteria do not effect students' evaluation for graduation.

TOPIC		COMMENTS
During the lesson did:		
Group members interact with each other?	Yes / No	
All group members participate?	Yes / No	
Group work as a team?	Yes / No	
Group members provide feedback?	Yes / No	
Each student publish during the ELC?	Yes / No	
Group adheres to the group rules?	Yes / No	
SDL interventions help the group?	Yes / No	
The group focuses on the task?	Yes / No	
Group achieves the training objective?	Yes / No	

Additional Comments:

SMALL GROUP INSTRUCTOR TRAINING COURSE

(SGITC)

TRAINING SUPPORT PACKAGE FOR

LESSON 7:

THE CONFERENCE

This page intentionally left blank.

Small Group Instruction

LESSON 7

THE CONFERENCE

L7-A1

Small Group Instruction

TERMINAL OBJECTIVE

Direct a conference so that participants gain:

- Insight into conference techniques.
- Synergy through consensus.
- Positive attitudes toward course content and its uses.

L7-A2

Small Group Instruction

ENABLING OBJECTIVE 1

Participate in a conference so all members:

- Contribute.
- Develop responses.
- Provide or solicit individual and group feedback.

Small Group Instruction

ENABLING OBJECTIVE 2

Develop group consensus by:

- Setting clear goals.
- Addressing problem causes.
- Seeking agreement.
- Supporting final decision.
- Using time efficiently.

Small Group Instruction

LESSON REVIEW

Direct a conference so that participants gain:

- **Insight into conference techniques.**
- **Synergy through consensus.**
- **Positive attitudes toward course content and its uses.**

L7-A5

NOT USED

LESSON 7

Student Handout

Handout	**Page**
Performance Evaluation Checklist	SG7-7

NOT USED

SGITC Lesson 7 – The Conference

Performance Evaluation Checklist

Purpose	The PEC is used to assess the SDL and the small group during the SDL presentation of an SGI lesson and final exam.
Who uses this checklist	The following personnel use this checklist: • Observing groups at the SDL site to rate the group and the SDL. • SGL to rate the SDL and group performance.
How to use this checklist	a. Observing groups: Fill out one checklist for the group. Discuss your observations and come to consensus. Ensure your discussion does not disturb the SDL's presentation. Look for positive and correctable SDL attributes and group dynamics. Try to link SDL roles, group process and content issues, and group dynamics to the success of the group meeting the training objective. Report out after the SDL reviews the ELC. b. SGL: Fill out checklist based on your observations of group and SDL performance.
When to use this checklist	This checklist is used for student lesson presentations (lessons 6-10) and the final exam.

NOT USED

SGITC · Lesson 7 – The Conference

PERFORMANCE EVALUATION CHECKLIST (Student Discussion Leader Rating)

Student Name: _____ Date: _____

Evaluator: _____ Rating: _____

Instructions: Evaluate student delivered lessons (6-10) and the final examination with this checklist. Students must receive the number of "GOs" listed next to the evaluated section, i.e., Introduction, 3 "GOs" out of 4. Students must pass each section (Introduction, Body, Conclusion, and ELC) to pass the course. Circle a GO or NO GO for each element in a section and the corresponding rating for that section. Write student rating (a GO or NO GO) on the Rating line above.

TOPIC		COMMENTS
Introduction (3 of 4). Did SDL:	**Go/ No Go**	
Focus group on task?	Go/ No Go	
Clearly state objective?	Go/ No Go	
Motivate/interest students?	Go/ No Go	
Tie SGI method to lesson objective?	Go/ No Go	
Lesson Body (3 of 3). Did SDL:	**Go/ No Go**	
Display SGL roles: - SME? - Facilitator? - Observer?	Go/ No Go	
Involve all students?	Go/ No Go	
Display understanding of SGI method chosen?	Go/ No Go	
ELC (4 of 4). Did SDL:	**Go/ No Go**	
Lead group in publishing stage?	Go/ No Go	
Lead group in processing stage?	Go/ No Go	
Lead group in generalizing stage?	Go/ No Go	
Lead group in applying stage?	Go/ No Go	
Conclusion (2 of 3). Did SDL:	**Go/ No Go**	
Review lesson objective and SGI method relationship?	Go/ No Go	
Summarize lesson results?	Go/ No Go	
Achieve training objective?	Go/ No Go	

Lesson 7 – The Conference   SGITC

PERFORMANCE EVALUATION CHECKLIST (Group)

Lesson: _____   Date: _____

SDL: _____   Evaluator: _____

Instructions: This checklist is used to evaluate group performance during student-led lessons (6-10). Observers within and outside the classroom use the checklist. The SGL uses this form to comment on group productivity and development.

Group criteria do not effect students' evaluation for graduation.

TOPIC		COMMENTS
During the lesson did:		
Group members interact with each other?	Yes / No	
All group members participate?	Yes / No	
Group work as a team?	Yes / No	
Group members provide feedback?	Yes / No	
Each student publish during the ELC?	Yes / No	
Group adheres to the group rules?	Yes / No	
SDL interventions help the group?	Yes / No	
The group focuses on the task?	Yes / No	
Group achieves the training objective?	Yes / No	

Additional Comments:

SMALL GROUP INSTRUCTOR TRAINING COURSE

(SGITC)

TRAINING SUPPORT PACKAGE FOR

LESSON 8:

ROLE PLAYING

This page intentionally left blank.

Small Group Instruction

LESSON 8
ROLE PLAYING

L8-A1

Small Group Instruction

TERMINAL OBJECTIVE

Conduct a role playing exercise so that students gain insight into role playing methodologies.

L8-A2

Small Group Instruction

PRACTICAL EXERCISE

L8-A3

Small Group Instruction

LESSON REVIEW

Conduct a role playing exercise so that students gain insight into role playing methodologies.

L8-A4

LESSON 8

Student Handouts and Exercises

Handout	**Page**
The Situation	SG8-7
Observer's Worksheet	SG8-9
Performance Evaluation Checklist	SG8-11

NOT USED

SGITC   Lesson 8 – Role Playing

The Situation

Situation	Widgex Systems has been a family-owned manufacturing company since 1945. Widgex supports a small southeastern community by providing assembly line and management jobs. The company prides itself on a high-quality widget. They attribute the consistently good product to the fair treatment of clients and employees, generous wages, and substantial employee benefits. Widgex purchases all of its raw materials within the United States and sells to small construction companies throughout the Southeast. In the last 53 years, the only real changes made in the assembly of the widgets have been machine upgrades, equipment replacements, and safety improvements.
The dilemma	For the last 5 years, the profit margin at Widgex has steadily dropped. Because of robotics and fewer trade barriers, it is now possible to obtain high-quality widgets at a reduced price including custom upgrades. Updating production methods and broadening the customer base appears to contradict the traditional company values that built its reputation and success.
The mission	There is a meeting of the Board to discuss the falling profit margin and company direction. Each of you must make your point of view known to the other members of the board. Everyone gets 2 minutes to state his or her case. Then the CEO opens the door floor discussion. At the end of 10-15 minutes, the CEO will make his announcements.
The players	The board members are:
	Robert Heartly, the President/CEO, and patriarch of the family.
	Albright Heartly, Vice President of Sales and oldest son of the family. He has been in charge of sales for about 7 years. His first year as Sales VP was a record year for Widgex. Albright is involved politically in the community.
	Emanuel Balancer, Vice President of Operations. He has held the job for 20 years.
	Pace Heartly, Vice President of Marketing. This is her second year on the job. She just graduated with honors from a well-respected business college in the Northeast.
	Victor Hernandez, Plant Foreman. A naturalized citizen from Mexico, with the company for 2 years. Victor has 2 brothers who run a widget plant in Mexico.
	Betty Winslow, Chief Accountant. Long time employee and high school girlfriend of Robert, the CEO. Very conservative financial officer, to the extreme.

NOT USED

SGITC   Lesson 8 – Role Playing

Observer's Worksheet

Your role   You are to observe the role playing exercise. Evaluate each role player using the worksheet below. Develop a consensus on what Widgex Corporation should do.

1. CEO's strategy, effectiveness, strong and weak points.
2. VP Sales strategy, effectiveness, strong and weak points.
3. VP of Operations strategy, effectiveness, strong and weak points.
4. Plant Foreman strategy, effectiveness, strong and weak points.
5. VP Marketing strategy, effectiveness, strong and weak points.
6. Chief Accountant strategy, effectiveness, strong and weak points.

NOT USED

Performance Evaluation Checklist

Purpose	The PEC is used to assess the SDL and the small group during the SDL presentation of an SGI lesson and final exam.
Who uses this checklist	The following personnel use this checklist: • Observing groups at the SDL site to rate the group and the SDL. • SGL to rate the SDL and group performance.
How to use this checklist	a. Observing groups: Fill out one checklist for the group. Discuss your observations and come to consensus. Ensure your discussion does not disturb the SDL's presentation. Look for positive and correctable SDL attributes and group dynamics. Try to link SDL roles, group process and content issues, and group dynamics to the success of the group meeting the training objective. Report out after the SDL reviews the ELC. b. SGL: Fill out checklist based on your observations of group and SDL performance.
When to use this checklist	This checklist is used for student lesson presentations (lessons 6-10) and the final exam.

NOT USED

SGITC   Lesson 8 – Role Playing

PERFORMANCE EVALUATION CHECKLIST (Student Discussion Leader Rating)

Student Name: _____   Date: _____

Evaluator: _____   Rating: _____

Instructions: Evaluate student delivered lessons (6-10) and the final examination with this checklist. Students must receive the number of "GOs" listed next to the evaluated section, i.e., Introduction, 3 "GOs" out of 4. Students must pass each section (Introduction, Body, Conclusion, and ELC) to pass the course. Circle a GO or NO GO for each element in a section and the corresponding rating for that section. Write student rating (a GO or NO GO) on the Rating line above.

TOPIC		COMMENTS
Introduction (3 of 4). Did SDL:	**Go/ No Go**	
Focus group on task?	Go/ No Go	
Clearly state objective?	Go/ No Go	
Motivate/interest students?	Go/ No Go	
Tie SGI method to lesson objective?	Go/ No Go	
Lesson Body (3 of 3). Did SDL:	**Go/ No Go**	
Display SGL roles: - SME? - Facilitator? - Observer?	Go/ No Go	
Involve all students?	Go/ No Go	
Display understanding of SGI method chosen?	Go/ No Go	
ELC (4 of 4). Did SDL:	**Go/ No Go**	
Lead group in publishing stage?	Go/ No Go	
Lead group in processing stage?	Go/ No Go	
Lead group in generalizing stage?	Go/ No Go	
Lead group in applying stage?	Go/ No Go	
Conclusion (2 of 3). Did SDL:	**Go/ No Go**	
Review lesson objective and SGI method relationship?	Go/ No Go	
Summarize lesson results?	Go/ No Go	
Achieve training objective?	Go/ No Go	

Lesson 8 – Role Playing   SGITC

PERFORMANCE EVALUATION CHECKLIST (Group)

Lesson: _____   Date: _____

SDL: _____   Evaluator: _____

Instructions: This checklist is used to evaluate group performance during student-led lessons (6-10). Observers within and outside the classroom use the checklist. The SGL uses this form to comment on group productivity and development.

Group criteria do not effect students' evaluation for graduation.

TOPIC		COMMENTS
During the lesson did:		
Group members interact with each other?	Yes / No	
All group members participate?	Yes / No	
Group work as a team?	Yes / No	
Group members provide feedback?	Yes / No	
Each student publish during the ELC?	Yes / No	
Group adheres to the group rules?	Yes / No	
SDL interventions help the group?	Yes / No	
The group focuses on the task?	Yes / No	
Group achieves the training objective?	Yes / No	

Additional Comments:

SMALL GROUP INSTRUCTOR TRAINING COURSE

(SGITC)

TRAINING SUPPORT PACKAGE FOR

LESSON 9:

COMMITTEE PROBLEM SOLVING

This page intentionally left blank.

Small Group Instruction

LESSON 9

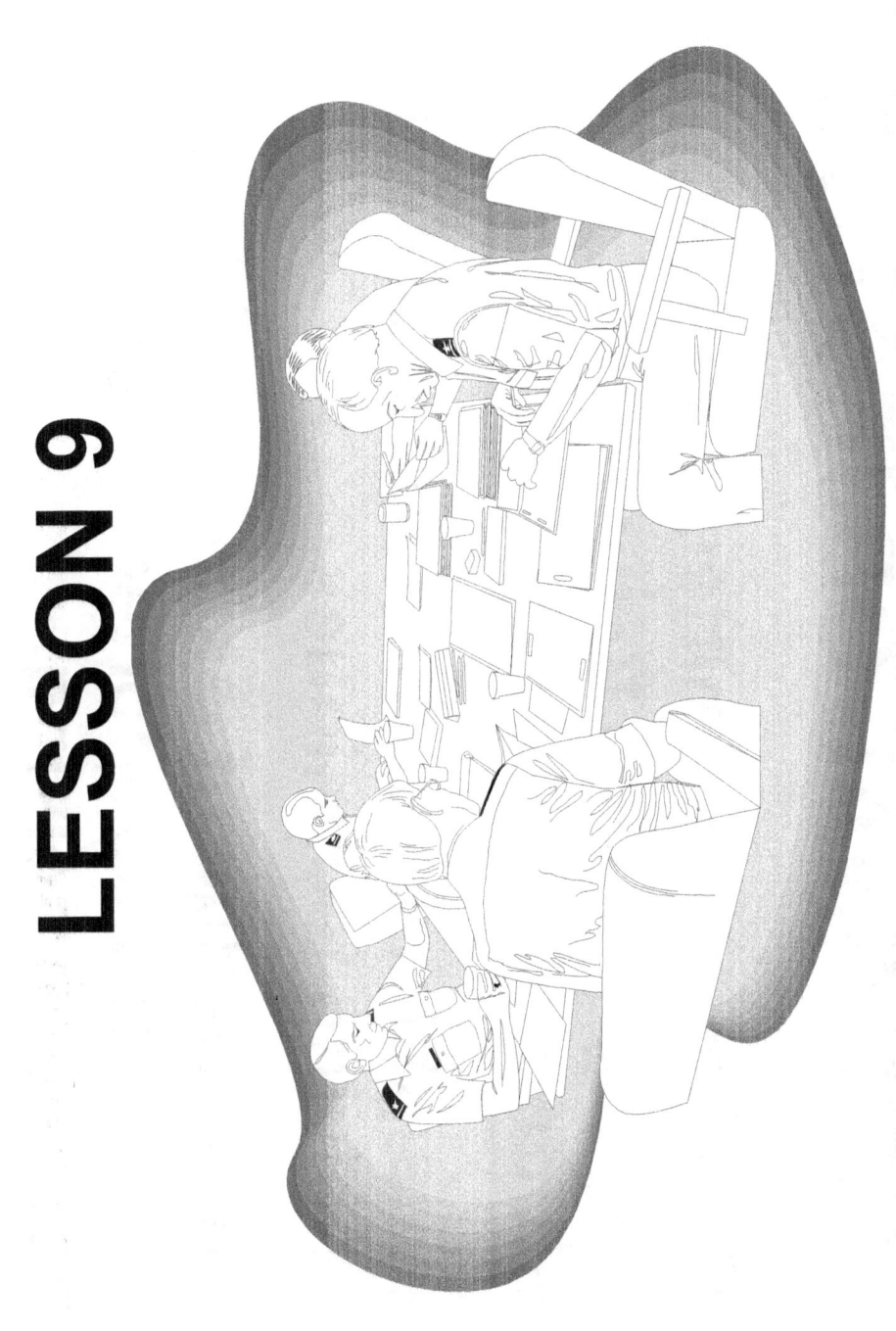

COMMITTEE PROBLEM SOLVING

L9-A1

Small Group Instruction

TERMINAL OBJECTIVE

Conduct a committee problem solving session to increase students':

- Insight into committee problem solving techniques.

- Knowledge of issues, problems, or topics supported by committee problem solving.

Small Group Instruction

ENABLING OBJECTIVE 1

Participate in committee problem solving by:

- **Collaborating in analysis and discussion of problem.**
- **Participating in fact finding.**
- **Obtaining group consensus.**
- **Providing or soliciting feedback.**

L9-A3

Small Group Instruction

ENABLING OBJECTIVE 2

Identify training objectives most likely achieved by committee problem solving.

L9-A4

Small Group Instruction

LESSON REVIEW

Conduct a committee problem solving session to increase students':

- Insight into committee problem solving techniques.

- Knowledge of issues, problems, or topics supported by committee problem solving.

NOT USED

LESSON 9

Student Handouts and Exercises

Exercise	**Page**
Committee Problem Solving-Training Development Exercise	SG9-9
Performance Evaluation Checklist	SG9-11

NOT USED

SGITC   Lesson 9 – Committee Problem Solving

Committee Problem Solving -- Training Development Exercise

Source This exercise was adapted from Pfeiffer, J. William, and Jones, John E., Ed. <u>A Handbook of Structured Experiences for Human Relations Training</u>. Pfeiffer and Co. (San Diego, CA: 1973).

Situation The new equipment will arrive at the school soon. The commandant wants more hands-on training in the courses. The director wants a plan to implement the commandant's desires. The analysis branch was tasked to review the training requirements. The new equipment training team was notified.

Requirement After reading the situation, reach a consensus within your small group on whether each of the statements is true (T), false (F), or unknown (?).

Statement	T	F	?
1. The plan must be finished before the equipment arrives.			
2. The training requirements are for the new equipment.			
3. The director wants a new equipment training plan.			
4. The director tasked the analysis branch.			
5. No one is reviewing the new equipment's training requirements.			
6. New equipment is scheduled to arrive at the school.			
7. The new equipment trainers will develop the plan after the analysis branch finishes.			
8. New equipment training trains the new equipment now.			
9. The revised training is for the new equipment.			
10. The following are true: the commandant wants more hands-on new equipment training and the director wants an implementation plan.			

NOT USED

SGITC   Lesson 9 – Committee Problem Solving

Performance Evaluation Checklist

Purpose	The PEC is used to assess the SDL and the small group during the SDL presentation of an SGI lesson and final exam.
Who uses this checklist	The following personnel use this checklist: • Observing groups at the SDL site to rate the group and the SDL. • SGL to rate the SDL and group performance.
How to use this checklist	a. Observing groups: Fill out one checklist for the group. Discuss your observations and come to consensus. Ensure your discussion does not disturb the SDL's presentation. Look for positive and correctable SDL attributes and group dynamics. Try to link SDL roles, group process and content issues, and group dynamics to the success of the group meeting the training objective. Report out after the SDL reviews the ELC. b. SGL: Fill out checklist based on your observations of group and SDL performance.
When to use this checklist	This checklist is used for student lesson presentations (lessons 6-10) and the final exam.

NOT USED

SGITC   Lesson 9 – Committee Problem Solving

PERFORMANCE EVALUATION CHECKLIST (Student Discussion Leader Rating)

Student Name: _____   Date: _____

Evaluator: _____   Rating: _____

Instructions: Evaluate student delivered lessons (6-10) and the final examination with this checklist. Students must receive the number of "GOs" listed next to the evaluated section, i.e., Introduction, 3 "GOs" out of 4. Students must pass each section (Introduction, Body, Conclusion, and ELC) to pass the course. Circle a GO or NO GO for each element in a section and the corresponding rating for that section. Write student rating (a GO or NO GO) on the Rating line above.

TOPIC		COMMENTS
Introduction (3 of 4). Did SDL:	**Go/ No Go**	
Focus group on task?	Go/ No Go	
Clearly state objective?	Go/ No Go	
Motivate/interest students?	Go/ No Go	
Tie SGI method to lesson objective?	Go/ No Go	
Lesson Body (3 of 3). Did SDL:	**Go/ No Go**	
Display SGL roles: - SME? - Facilitator? - Observer?	Go/ No Go	
Involve all students?	Go/ No Go	
Display understanding of SGI method chosen?	Go/ No Go	
ELC (4 of 4). Did SDL:	**Go/ No Go**	
Lead group in publishing stage?	Go/ No Go	
Lead group in processing stage?	Go/ No Go	
Lead group in generalizing stage?	Go/ No Go	
Lead group in applying stage?	Go/ No Go	
Conclusion (2 of 3). Did SDL:	**Go/ No Go**	
Review lesson objective and SGI method relationship?	Go/ No Go	
Summarize lesson results?	Go/ No Go	
Achieve training objective?	Go/ No Go	

Lesson 9 – Committee Problem Solving   SGITC

PERFORMANCE EVALUATION CHECKLIST (Group)

Lesson: _____   Date: _____

SDL: _____   Evaluator: _____

Instructions: This checklist is used to evaluate group performance during student-led lessons (6-10). Observers within and outside the classroom use the checklist. The SGL uses this form to comment on group productivity and development.

Group criteria do not effect students' evaluation for graduation.

TOPIC		COMMENTS
During the lesson did:		
Group members interact with each other?	Yes / No	
All group members participate?	Yes / No	
Group work as a team?	Yes / No	
Group members provide feedback?	Yes / No	
Each student publish during the ELC?	Yes / No	
Group adheres to the group rules?	Yes / No	
SDL interventions help the group?	Yes / No	
The group focuses on the task?	Yes / No	
Group achieves the training objective?	Yes / No	

Additional Comments:

SMALL GROUP INSTRUCTOR TRAINING COURSE

(SGITC)

TRAINING SUPPORT PACKAGE FOR

LESSON 10:

CASE STUDIES

This page intentionally left blank.

Small Group Instruction

LESSON 10
CASE STUDIES

CASE STUDIES:
Abbreviated Printed

Small Group Instruction

L10a-A2

Small Group Instruction

TERMINAL OBJECTIVE

Lead an abbreviated printed case discussion so students:

- Identify issues and problems.
- Identify diverse viewpoints on single issue or problem.
- Recognize underlying principles.
- Apply identified principles to problem diagnosis and solution.

Small Group Instruction

ENABLING OBJECTIVE 1

Participate in an abbreviated printed case discussion by:

- Identifying issues and problems.
- Identifying underlying principles.
- Participating in case analysis and discussion.
- Developing possible solutions.
- Providing or soliciting feedback.

Small Group Instruction

ENABLING OBJECTIVE 2

Develop group solution by:

- Collaborating to analyze case.
- Applying case principles.
- Obtaining member consensus.

Small Group Instruction

LESSON REVIEW

Lead an abbreviated printed case discussion so students:

- Identify issues and problems.
- Identify diverse viewpoints on single issue or problem.
- Recognize underlying principles.
- Apply identified principles to problem diagnosis and solution.

LESSON 10a

Student Handouts and Exercises

Handout	**Page**
Basic Training Instructions: Group Discussion	SG10a-9
Basic Training Reading	SG10a-11
Performance Evaluation Checklist	SG10a-13

NOT USED

SGITC   Lesson 10a – Abbreviated Printed Case Discussion

Basic Training Instructions: Group Discussion

Instructions   Break into small groups. Read the case and discuss the following issues. Be prepared to present your rationale for your group answers.

1. What should the squad leader do?

2. What should the drill sergeant do?

3. What should the squad members do?

4. What type of leadership should be used by each?

5. Discuss the type of interventions that could be used, and when they should be used.

6. Compare the situation to SGI.

NOT USED

SGITC   Lesson 10a – Abbreviated Printed Case Discussion

Basic Training Reading

Situation

A group of young men from the Northeast arrived at a reception center to start basic training. They were nervous and unsure of what was going to happen to them. They had heard stories of mean Drill Sergeants, gas chambers, and bad food. They also heard of having to crawl in mud and water through barbed wire while machine guns were being fired. They were a little intimidated.

As they began to process in, several trainees began to become friends. Some discovered that they were from the same hometown or liked a particular type of music or had other similar interests. These relationships began to grow as most of the men were assigned to the same company, and ultimately, the same platoon.

After in processing, the fun started fast and furiously. It was the middle of the first night when all were called to attention in front of the barracks in full gear. Everyone wondered what was going on. Apparently, someone in another platoon had lost something of value and the search was on to find it. It did not turn up. So everyone ran around in formation screaming silly things and holding their hands over their heads as they ran along. When they stopped running it was time for breakfast.

The second and third days were equally pleasant. They missed lunch on the second day because one of the trainees could not hit the target with his rifle. One of the squad members told the trainee to throw the bullet at it so everyone could go to lunch. The trainee did, and soon everyone was down range looking for the bullet. Bye-bye lunch. The third day started with a fire-drill at zero-dark-thirty and ended with a real fire in the dining facility.

The trainees were still nervous and curious about the rest of the training, but there wasn't any time to devote to these feelings; too much was going on. They knew, however, that there was an exercise coming up that everyone was talking about, and it seemed very important. Something about night-fire and barbed wire.

As the days grew longer and the training more intense, several members of one of the squads became disenchanted with their situation. They began to complain and become uncooperative. They did not pay attention in class and did a minimum of physical training to get by. They found that they could slack-off a little and have the rest of the squad make up the difference. It annoyed some, but time was precious and the other squad members had their hands full doing their own jobs.

Continued on next page

Lesson 10a – Abbreviated Printed Case Discussion

Basic Training Reading, Continued

The training continued as basic training will, and the unhappy trainees' attitudes and performance became apparent to the squad leader and the drill sergeant. However, no one spoke to the few malcontents, and the situation continued. Oh, there were a few squad members who would make small talk about the lack of teamwork from everyone or that the lack of effort by some was making everyone else work harder. But there was no change in behavior.

The squad leader was concerned because he knew that an upcoming exercise was required the squad to function like a well-oiled machine. Currently, it needed a tune-up. The squad leader had a problem.

The drill sergeant for the squad was in the PX barber shop the weekend before the night fire exercise. As he was sitting in the barber's chair, a drill sergeant from another platoon came up to him and mentioned that the exercise was coming up. He said his squads could beat any other squads at any game, at any time. He went on to mention that he had noticed some slackers around, and, though he did not want to mention any names, he might be looking at someone who had some "problems" with his troops. On his way out, he wished the drill luck; he thought he was going to need it.

While getting his hair cut, the drill sergeant reviewed in his mind the past performance of his squads. He was well aware of each man's capabilities and knew how each man and squad keyed off of the talent and teamwork of the other. He knew that one of his squads was not working to its potential. He also knew that he did not need any "luck" to beat another platoon, squad by squad. No one was going to put his unit down. It was the third week of training, and the drill sergeant had a problem.

Performance Evaluation Checklist

Purpose	The PEC is used to assess the SDL and the small group during the SDL presentation of an SGI lesson and final exam.
Who uses this checklist	The following personnel use this checklist: • Observing groups at the SDL site to rate the group and the SDL. • SGL to rate the SDL and group performance.
How to use this checklist	a. Observing groups: Fill out one checklist for the group. Discuss your observations and come to consensus. Ensure your discussion does not disturb the SDL's presentation. Look for positive and correctable SDL attributes and group dynamics. Try to link SDL roles, group process and content issues, and group dynamics to the success of the group meeting the training objective. Report out after the SDL reviews the ELC. b. SGL: Fill out checklist based on your observations of group and SDL performance.
When to use this checklist	This checklist is used for student lesson presentations (lessons 6-10) and the final exam.

Lesson 10a – Abbreviated Printed Case Discussion   SGITC

NOT USED

SGITC   Lesson 10a – Abbreviated Printed Case Discussion

PERFORMANCE EVALUATION CHECKLIST (Student Discussion Leader Rating)

Student Name: _____   Date: _____

Evaluator: _____   Rating: _____

Instructions: Evaluate student delivered lessons (6-10) and the final examination with this checklist. Students must receive the number of "GOs" listed next to the evaluated section, i.e., Introduction, 3 "GOs" out of 4. Students must pass each section (Introduction, Body, Conclusion, and ELC) to pass the course. Circle a GO or NO GO for each element in a section and the corresponding rating for that section. Write student rating (a GO or NO GO) on the Rating line above.

TOPIC		COMMENTS
Introduction (3 of 4). Did SDL:	**Go/ No Go**	
Focus group on task?	Go/ No Go	
Clearly state objective?	Go/ No Go	
Motivate/interest students?	Go/ No Go	
Tie SGI method to lesson objective?	Go/ No Go	
Lesson Body (3 of 3). Did SDL:	**Go/ No Go**	
Display SGL roles: - SME? - Facilitator? - Observer?	Go/ No Go	
Involve all students?	Go/ No Go	
Display understanding of SGI method chosen?	Go/ No Go	
ELC (4 of 4). Did SDL:	**Go/ No Go**	
Lead group in publishing stage?	Go/ No Go	
Lead group in processing stage?	Go/ No Go	
Lead group in generalizing stage?	Go/ No Go	
Lead group in applying stage?	Go/ No Go	
Conclusion (2 of 3). Did SDL:	**Go/ No Go**	
Review lesson objective and SGI method relationship?	Go/ No Go	
Summarize lesson results?	Go/ No Go	
Achieve training objective?	Go/ No Go	

Lesson 10a – Abbreviated Printed Case Discussion   SGITC

PERFORMANCE EVALUATION CHECKLIST (Group)

Lesson: _____   Date: _____

SDL: _____   Evaluator: _____

Instructions: This checklist is used to evaluate group performance during student-led lessons (6-10). Observers within and outside the classroom use the checklist. The SGL uses this form to comment on group productivity and development.

Group criteria do not effect students' evaluation for graduation.

TOPIC		COMMENTS
During the lesson did:		
Group members interact with each other?	Yes / No	
All group members participate?	Yes / No	
Group work as a team?	Yes / No	
Group members provide feedback?	Yes / No	
Each student publish during the ELC?	Yes / No	
Group adheres to the group rules?	Yes / No	
SDL interventions help the group?	Yes / No	
The group focuses on the task?	Yes / No	
Group achieves the training objective?	Yes / No	

Additional Comments:

Small Group Instruction

LESSON 10
CASE STUDIES

L10b-A1

CASE STUDIES:
Dramatized

Small Group Instruction

L10b-A2

Small Group Instruction

TERMINAL OBJECTIVE

Lead an abbreviated dramatized case discussion so students:

- Identify issues and problems.
- Identify diverse viewpoints on single issue or problem.
- Recognize underlying principles.
- Apply identified principles to problem diagnosis and solution.

Small Group Instruction

ENABLING OBJECTIVE 1

Participate in an abbreviated dramatized case discussion by:

- Identifying issues and problems.
- Identifying underlying principles.
- Participating in case analysis and discussion.
- Developing possible solutions.
- Providing or soliciting feedback.

L10b-A4

Small Group Instruction

ENABLING OBJECTIVE 2

Develop group solution by:

- Collaborating to analyze case.
- Applying case principles.
- Selecting two options.

Small Group Instruction

LESSON REVIEW

Lead an abbreviated dramatized case discussion so students:

- Identify issues and problems.
- Identify diverse viewpoints on single issue or problem.
- Recognize underlying principles.
- Apply identified principles to problem diagnosis and solution.

SGITC Lesson 10b – Abbreviated Dramatized Case Discussion

LESSON 10b

Student Handouts and Exercises

Handout	**Page**
Performance Evaluation Checklist	SG10a-9

NOT USED

SGITC   Lesson 10b – Abbreviated Dramatized Case Discussion

Performance Evaluation Checklist

Purpose	The PEC is used to assess the SDL and the small group during the SDL presentation of an SGI lesson and final exam.
Who uses this checklist	The following personnel use this checklist: • Observing groups at the SDL site to rate the group and the SDL. • SGL to rate the SDL and group performance.
How to use this checklist	a. Observing groups: Fill out one checklist for the group. Discuss your observations and come to consensus. Ensure your discussion does not disturb the SDL's presentation. Look for positive and correctable SDL attributes and group dynamics. Try to link SDL roles, group process and content issues, and group dynamics to the success of the group meeting the training objective. Report out after the SDL reviews the ELC. b. SGL: Fill out checklist based on your observations of group and SDL performance.
When to use this checklist	This checklist is used for student lesson presentations (lessons 6-10) and the final exam.

NOT USED

SGITC Lesson 10b – Abbreviated Dramatized Case Discussion

PERFORMANCE EVALUATION CHECKLIST (Student Discussion Leader Rating)

Student Name: _____   Date: _____

Evaluator: _____   Rating: _____

Instructions: Evaluate student delivered lessons (6-10) and the final examination with this checklist. Students must receive the number of "GOs" listed next to the evaluated section, i.e., Introduction, 3 "GOs" out of 4. Students must pass each section (Introduction, Body, Conclusion, and ELC) to pass the course. Circle a GO or NO GO for each element in a section and the corresponding rating for that section. Write student rating (a GO or NO GO) on the Rating line above.

TOPIC		COMMENTS
Introduction (3 of 4). Did SDL:	**Go/ No Go**	
Focus group on task?	Go/ No Go	
Clearly state objective?	Go/ No Go	
Motivate/interest students?	Go/ No Go	
Tie SGI method to lesson objective?	Go/ No Go	
Lesson Body (3 of 3). Did SDL:	**Go/ No Go**	
Display SGL roles: - SME? - Facilitator? - Observer?	Go/ No Go	
Involve all students?	Go/ No Go	
Display understanding of SGI method chosen?	Go/ No Go	
ELC (4 of 4). Did SDL:	**Go/ No Go**	
Lead group in publishing stage?	Go/ No Go	
Lead group in processing stage?	Go/ No Go	
Lead group in generalizing stage?	Go/ No Go	
Lead group in applying stage?	Go/ No Go	
Conclusion (2 of 3). Did SDL:	**Go/ No Go**	
Review lesson objective and SGI method relationship?	Go/ No Go	
Summarize lesson results?	Go/ No Go	
Achieve training objective?	Go/ No Go	

Lesson 10b – Abbreviated Dramatized Case Discussion   SGITC

PERFORMANCE EVALUATION CHECKLIST (Group)

Lesson: _____   Date: _____

SDL: _____   Evaluator: _____

Instructions: This checklist is used to evaluate group performance during student-led lessons (6-10). Observers within and outside the classroom use the checklist. The SGL uses this form to comment on group productivity and development.

Group criteria do not effect students' evaluation for graduation.

TOPIC		COMMENTS
During the lesson did:		
Group members interact with each other?	Yes / No	
All group members participate?	Yes / No	
Group work as a team?	Yes / No	
Group members provide feedback?	Yes / No	
Each student publish during the ELC?	Yes / No	
Group adheres to the group rules?	Yes / No	
SDL interventions help the group?	Yes / No	
The group focuses on the task?	Yes / No	
Group achieves the training objective?	Yes / No	

Additional Comments:

Small Group Instruction

LESSON 10
CASE STUDIES

L10c-A1

CASE STUDIES:
Incident-Process

Small Group Instruction

L10c-A2

Small Group Instruction

TERMINAL OBJECTIVE

Lead an incident-process case discussion so students gain:

- **Recognition of issues and problems.**
- **Problem-solving skills.**
- **Fact-finding skills.**

L10c-A3

Small Group Instruction

ENABLING OBJECTIVE 1

Participate in an incident-process case discussion by:

- Identifying required information.
- Participating in fact finding.
- Providing or soliciting feedback.
- Developing group solution.

Small Group Instruction

ENABLING OBJECTIVE 2

Develop group solution by:

- Collaborating to analyze case.
- Seeking additional information relevant to problem.
- Obtaining member consensus.
- Assessing adequacy of final decision.

Small Group Instruction

LESSON REVIEW

Lead and incident-process case discussion so students gain:

- **Recognition of issues and problems.**
- **Problem-solving skills.**
- **Fact-finding skills.**

L10c-A6

SGITC Lesson 10c – Incident Process Case Discussion

LESSON 10c

Student Handouts and Exercises

Handout	**Page**
Team Instruction Sheet	SG10c-9
Performance Evaluation Checklist	SG10c-11

NOT USED

SGITC Lesson 10c – Incident Process Case Discussion

Team Instruction Sheet

Situation

a. Your team is working on a project at battalion headquarters. The team is planning a "military stakes" exercise where soldiers rotate through stations and are evaluated on their ability to perform common tasks. A noncommissioned officer in charge (NCOIC) has been assigned for each station. The stations will be located in the field adjacent to the headquarters building.

b. The operations officer has talked with team members individually about the plan, what has been done so far, and the remaining tasks to be accomplished. Unfortunately, no one team member seems to have all the required information needed to complete the plan. Further, the operations officer has just been called away from the office unexpectedly. Before departing, the operations officer announced to everyone, "I want you to complete the plan while I'm gone. I'll expect the final answers when I return in 15 minutes."

Requirement

Here is what must be done:
- Your team may begin work when all of its members have finished reading these instructions.
- Each team member will receive written bits of information. These are <u>not</u> to be shown to other team members.
- What will be required, and how to go about it, will become clear as you <u>share</u> information with the other team members through <u>verbal communications only</u>.
- If, after sharing information with other members, you still feel you lack relevant information, ask specific questions of your facilitator.
- When you and your team members feel that the team has completed the required tasks, call on your facilitator to check your results.
- If you have only partially completed your tasks, or if you have done more than what was required, your facilitator will consider the tasks totally incomplete. In that case, you will have to keep working without knowing which part of your task, if any, has been completed satisfactorily.

Continued on next page

Lesson 10c – Incident Process Case Discussion

Team Instruction Sheet, Continued

Rules

Here is a list of rules that you must follow:

- From the moment your team begins work, you may speak only to other team members and the team's facilitator.
- You may not show others the contents of your written bits of information.
- If you ask your team's facilitator for more information, then --

  - You must be specific. There will be no response to general requests for more information.
  - Do not expect information to be volunteered unless specifically requested.
  - There will be no speculation about information that is not available.

- You must obey the facilitator's instructions.
- Your team's work must be completed in 45 minutes.

Performance Evaluation Checklist

Purpose	The PEC is used to assess the SDL and the small group during the SDL presentation of an SGI lesson and final exam.
Who uses this checklist	The following personnel use this checklist: • Observing groups at the SDL site to rate the group and the SDL. • SGL to rate the SDL and group performance.
How to use this checklist	a. Observing groups: Fill out one checklist for the group. Discuss your observations and come to consensus. Ensure your discussion does not disturb the SDL's presentation. Look for positive and correctable SDL attributes and group dynamics. Try to link SDL roles, group process and content issues, and group dynamics to the success of the group meeting the training objective. Report out after the SDL reviews the ELC. b. SGL: Fill out checklist based on your observations of group and SDL performance.
When to use this checklist	This checklist is used for student lesson presentations (lessons 6-10) and the final exam.

NOT USED

SGITC   Lesson 10c – Incident Process Case Discussion

PERFORMANCE EVALUATION CHECKLIST (Student Discussion Leader Rating)

Student Name: _____   Date: _____

Evaluator: _____   Rating: _____

Instructions: Evaluate student delivered lessons (6-10) and the final examination with this checklist. Students must receive the number of "GOs" listed next to the evaluated section, i.e., Introduction, 3 "GOs" out of 4. Students must pass each section (Introduction, Body, Conclusion, and ELC) to pass the course. Circle a GO or NO GO for each element in a section and the corresponding rating for that section. Write student rating (a GO or NO GO) on the Rating line above.

TOPIC		COMMENTS
Introduction (3 of 4). Did SDL:	**Go/ No Go**	
Focus group on task?	Go/ No Go	
Clearly state objective?	Go/ No Go	
Motivate/interest students?	Go/ No Go	
Tie SGI method to lesson objective?	Go/ No Go	
Lesson Body (3 of 3). Did SDL:	**Go/ No Go**	
Display SGL roles: - SME? - Facilitator? - Observer?	Go/ No Go	
Involve all students?	Go/ No Go	
Display understanding of SGI method chosen?	Go/ No Go	
ELC (4 of 4). Did SDL:	**Go/ No Go**	
Lead group in publishing stage?	Go/ No Go	
Lead group in processing stage?	Go/ No Go	
Lead group in generalizing stage?	Go/ No Go	
Lead group in applying stage?	Go/ No Go	
Conclusion (2 of 3). Did SDL:	**Go/ No Go**	
Review lesson objective and SGI method relationship?	Go/ No Go	
Summarize lesson results?	Go/ No Go	
Achieve training objective?	Go/ No Go	

Lesson 10c – Incident Process Case DiscussionSGITC

PERFORMANCE EVALUATION CHECKLIST (Group)

Lesson: _____ Date: _____

SDL: _____ Evaluator: _____

Instructions: This checklist is used to evaluate group performance during student-led lessons (6-10). Observers within and outside the classroom use the checklist. The SGL uses this form to comment on group productivity and development.

Group criteria do not effect students' evaluation for graduation.

TOPIC		COMMENTS
During the lesson did:		
Group members interact with each other?	Yes / No	
All group members participate?	Yes / No	
Group work as a team?	Yes / No	
Group members provide feedback?	Yes / No	
Each student publish during the ELC?	Yes / No	
Group adheres to the group rules?	Yes / No	
SDL interventions help the group?	Yes / No	
The group focuses on the task?	Yes / No	
Group achieves the training objective?	Yes / No	

Additional Comments:

This page intentionally left blank.

www.ingramcontent.com/pod-product-compliance
Lightning Source LLC
Chambersburg PA
CBHW060310240426
43661CB00059B/2712